Stepping Stones

APPROPRIATE TECHNOLOGY
AND BEYOND

edited by Lane de Moll and Gigi Coe

Schocken Books • New York

Editors   Lane deMoll and Gigi Coe
Layout and Design   Linda Sawaya
Typesetting   Irish Setter

With thanks to the people at Rainhouse and the California Office of Appropriate Technology for their good ideas and moral support. Thanks also to all the people who have given us permission to reprint their material.

The cover drawing by Diane Schatz is available as a poster for $3 from *Rain,* 2270 N.W. Irving, Portland, OR 97210.

Where to Go for More Information

The most basic resources in appropriate technology are referenced throughout the text via bracketed numbers which refer to the resource listings beginning on page *198*. For a more thorough compilation with evaluations of the broad range of projects, books and research projects in the field, one should turn to *Rainbook: Resources in Appropriate Technology,* 1977, $7.95 from Schocken Books, 200 Madison Avenue, New York, NY, or from *Rain*.

Rain: Journal of Appropriate Technology is a monthly magazine providing continuing information access and discussion on the issues and ideas presented in *Stepping Stones. Rain* is available for $15/year ($7.50/year living lightly rate) from 2270 N.W. Irving, Portland, OR 97210. Send $1 for sample copy.

First published by SCHOCKEN BOOKS 1978

10 9 8 7 6 5 4 3 2 1   78 79 80 81

Copyright © 1978 by Schocken Books Inc.

Library of Congress Cataloging in Publication Data
    Main entry under title:

    Stepping stones.

    1. Technology—Addresses, essays, lectures. 2. Technology—Social aspects—Addresses, essays, lectures. 3. Human ecology—Addresses, essays, lectures. I. De Moll, Lane. II. Coe, Gigi.

T185.S85   608   78-54392

Manufactured in the United States of America

TOM BENDER

In Japan, the stepping stones in a garden are spaced far enough apart so that each step must be taken carefully, one by one. As you look down to cross safely, you see in the water the reflections of the clouds and trees above. Your attention is focussed from the broad view of the garden to the detail of the immediate footstep and the mirror of the sky . . . When your footing is again secure you look up to find yourself in a new place.

TABLE OF CONTENTS

INTRODUCTION

The paths by which ideas take seed and grow are often subtle. Although specific steps may vary with different individuals, common threads of experience nurture ideas and convictions. These become experiments which in turn stimulate ideas that are enlarged upon as they are passed around. New skills and resources emerge as well.

The ideas of appropriate technology grew out of a shared gut level sense that something somehow was seriously wrong with our way of doing things. People were becoming increasingly disenchanted with a way of life that allowed the squandering of natural resources—and money—for questionable material gains. Outrage against a senseless war in a tiny country in Asia pushed many people into action seeking a less violent alternative future. Others reacted to the rapidly deteriorating quality of the physical, social and economic environment—polluted air and water, unfulfilling jobs and sprawling urbanization—by developing more meaningful and equitable life-styles and communities.

Those who responded by going "back to the land" to explore the possibilities and value of self-sufficiency became an important source of innovations in alternative agriculture and the use of the sun, wind, water and biofuels for clean energy. Others honed their political and organizing skills working for human rights and economic justice. Most recently people in small towns and urban neighborhoods have also begun reaching out for alternative solutions to their problems. These groups are now coming together to form a social and political force which has the potential to change the course of our society.

The pieces collected here are the philosophical stepping stones which have helped shape the techniques, values, tools and politics of appropriate technology. The book arose out of a need we both felt in our work with Rain Magazine and the California Office of Appropriate Technology for something which pulled together and clarified the most important values and ideas to come out of this exploration. It is intended as a companion to Rainbook: Resources in Appropriate Technology, which documents the manifestations of the ideas presented here.

The first part—The Party's Over—focusses on the early work which defined the importance of and need for change. Part Two elaborates some qualities of an emerging enabling technology that is more appropriate and relevant to human needs than the path our society has been pursuing.

The pieces in these two sections have appeared before in a variety of places. Most have been reprinted several times in relatively obscure "alternative" journals like Manas, Resurgence, Co-Evolution Quarterly, and Rain. Although increasingly they are appearing in such established places as Foreign Affairs and Smithsonian. Many were circulated hand to hand through a network via tattered xerox and mimeographed copies.

The final section of the book is both a reflection on the implications of what has happened to date and an exploration of what needs yet to be done. The last five pieces were written especially for this book by David Morris, Lee Johnson, Gil Friend, Margaret Mead and Tom Bender in response to a question about what lies beyond "appropriate technology." Use them as stepping stones and a reminder that change is possible—not only possible, but happening. □

2

PART I

4

THE PARTY'S OVER

The discovery of rich sources of petroleum in the mid-1800s began an era of wealth which allowed us to test our ingenuity to the limit—to run the course of our imaginations without the constraints of efficiency or frugality to slow us down. During that time we reached for the moon—and got it.

Every era has had its frontiers of new worlds to conquer, and each has consumed a lot in the conquering. The hunger for tall, straight trees for masts, land to farm and whale oil for lamps drains resources and taxes the capacity of natural systems to maintain themselves and take care of us at the same time.

Even before oil was discovered, when coal fueled progress, discerning people saw what was happening around them. Go to any library and look for a book titled Man and Nature. It probably hasn't been checked out for years. In 1878, when our frontiers seemed endless and newly acquired Alaska was being hailed as the Louisiana Purchase of the 19th century, George Marsh wrote this book protesting the ruthless destruction of the land:

> The earth is fast becoming an unfit home for its noblest inhabitant, and another era of human crime and human improvidence, and of like duration . . . would reduce it to such a condition of impoverished productiveness, of shattered surface, of climatic excess, as to threaten the depravation, barbarism, and perhaps even extinction of the species. (39)

His warning and those of others like him and previous civilizations were not heeded. Questions of long-range survival were never asked in the rush to monopolize the apparently unlimited resources.

In 1922, A. J. Lotka tried to explain survival in this way:

> . . . the fundamental object of contention in the life-struggle, in the evolution of the organic world, is available energy. . . . In the struggle for existence the advantage must go to those organisms whose energy capturing devices are most efficient in directing available energy into channels favorable to the preservation of the species.
>
> The first effect in natural selection thus operating on competing species will be to give relative preponderance (in numbers or mass) to those most efficient in guiding available energy in the manner indicated (towards survival.)(39)

In a period when there are excess, unclaimed amounts of energy available from the sun, soil, water and nutrients in the system, the ones who can capture that energy most quickly win out. But when energy sources are limited, survival goes to those who are most able to adapt to changing conditions. The definition of efficiency alters and the emphasis changes from quantity to quality.

Thus, in a clearing on the Oregon coast, the alder shoot up first, being the most able to capture quickly the sunlight and space available. Under cover of their shade and with the help of the nutrients they put into the soil, the tiny spruce trees start. They grow slowly and carefully, needing protection and nourishment from the "weed" trees. As the spruce eventually reach their maturity, they shade out the alder, who have already begun to die—blowing their tops out in the wind because their quick growth has not given them the strength to endure. They rot from the inside—leaving the area to the spruce who grow strong enough to stand firm in the full force of the weather hitting the west coast of the continent.

The analogy is often made between the plant world and humans. When there was cheap energy to burn, we used it—growing quickly and powerfully to dominate the world. The fossil fuel era isn't the first time this has happened. Land and resources throughout the ages have been exploited. Even the Chinese, whose frugality we now admire, are following an ancient cycle of exploitation and regeneration. Successive generations have had to plant new trees after previous ones used up the valuable forests without a thought for the future. As conditions change, so must the cycles of cultures and species. Recently we have been mostly alders. Now we must cultivate the spruce among us.

As we enter a time when resources are becoming more scarce, we have the option of muddling through making changes as they become unavoidable, or we can act now using the knowledge, tools and experience we have to make a relatively comfortable transition. □

C. R. Ashbee wrote this piece in 1909 as the forward to The Indian Craftsman *by Ananda K. Coomaraswamy (17), which describes the history and customs of the craft guilds in the villages of India. Putting India's crafts in perspective, Ashbee looks at the effect of industrialization on our crafts, our communities and our souls.* The Indian Craftsman *is now out of print but might be found in the library.*

SPEAK TO THE EARTH & IT SHALL TEACH THEE

By C. R. Ashbee

IF WE EXAMINE our own Western economic history, more particularly the history of England, we find that the break-up of the conditions of English craftsmanship and the English village order cannot be traced back beyond the industrial revolution of the 18th century, and the enclosure of the common lands that accompanied it. Fundamentally, with us the great change came with the introduction of industrial machinery, and the question which forces itself upon us is: What are the benefits to our culture of the industrial machinery that has acted in this manner?

Trained as we are to measure everything by a mechanical standard, it is difficult for us to see things clearly, to get a correct focus. We are apt to forget that our view is biassed, that we attach a disproportionate value to the productions of machinery, and that a vast number, perhaps 60 percent, of these productions are not, as is generally supposed, labor-saving, health-giving and serviceable to our general life and culture, but the reverse. "It is questionable," said John Stuart Mill, half a century ago, "whether all the labor-saving machinery has yet lightened the day's labor of a single human being;" and the years that have followed his death seem not only to have further borne out his statement, but the

people themselves who are being exploited by mechanical conditions are beginning to find it out.

For machinery is only a measure of human force, not an increase of it; and it is questionable whether, owing to the abuse of machinery, the destruction and waste it brings may not equal the gain it yields. Wonderful are the great ships, and the winged words from one side of the globe to the other, wonderful is the consciousness that comes to us from those things, and from rapid movement, and from our power of destruction, but we may pay even too high a price for the boon of progress. It behooves us to ask, at least, what the price is, and if it be a fair one. Perchance, in our thoughtlessness we have, like the boy in the fairy tale, bartered away the cow for a handful of beans; well, there may be much virtue in the beans, but was not the cow good too? A more reasonable view of life and the progress of Western civilization is making us see that the pitiful slums of our great cities are not a necessary corollary to the great ships; that a nicer, saner regulation of industry will mean that the rapid displacement of human labor, and the misery it brings, may be graduated and softened; that it is not necessary for 30 percent of the population to die in pauperism, as is the case in England at present; that it is short-sighted and unwise to paralyze invention and skill and individuality by unregulated machine development, and that our present gauge of the excellence in all these things—their saleability—cannot possibly continue to be a permanent gauge.

It is when Western civilization is brought face to face with the results of other cultures, Eastern cultures, when the stages of its progress are resumed from the points of view of other religions, when Japan, for instance, rejects or chooses what she needs to make her a fighting force, when India seeks to form, out of an imposed educational system, a political consciousness of her own; when Persia and Turkey are in the act of creating constitutions on the Western model, that we in the West come to realize that the stages of progress, as we understand them, are not obligatory. Some of them may be skipped.

So it is with industry, and the conditions of life induced by industrial machinery, much of it may be skipped. And it is the continuous existence of an order like the Indian village community which, when brought into relation with Western progress, seems to prove this. We in the West have passed through the condition of the Aryan village community. The conditions in India and Ceylon are very similar to the conditions that prevailed in medieval England, Germany and France; they did not seem nearly so strange to Knox, writing in Stuart times, as they do to us. The 500 years that have passed between our middle ages and the growth of the great cities of machine industry may have proved that the destruction of the Western village community was inevitable, but it has not proved that where the village community still exists it need necessarily be destroyed. Indeed, we are finding out in the West that if the village tradition were still living it could still be utilized; we are even seeking to set up something like it in its place. For the great city of mechanical industry has come to a point when its disintegration is inevitable. There are signs that the devolution has already begun, both in England and America. The cry of "back to the land," the plea for a "more reasonable life," the revival of the handicrafts, the education of hand and eye, the agricultural revival, the German *ackerbau*," the English small holding, our technical schools, all these things are indications of a need for finding something, if not to take the place of the village community, at least to bring once again into life those direct, simple, human and out-of-door things of which mechanical industry has deprived our working popu-

lation. These things are necessary to our life as a people, and we shall have to find them somehow. Dr. Coomaraswamy does well to show how they still exist in great measure in the East, and it may be that the East, in her wisdom, and with her profound conservative instinct, will not allow them to be destroyed. She has, as Sir Geo. Birdwood puts it, let the races and the peoples for 3,000 years come and pass by; she may have taken this from one and that from another, but the fundamental democratic order of her society has remained, and it appears improbable, on the face of it, that we English shall materially change it when so many others before us have left it undisturbed.

Indeed, there seems to be no reason, on the face of it, why we should aspire to do so. Some change we are certainly bringing, and bringing unconsciously, but it is a curious and suggestive thought that the spiritual reawakening in England, which goes now by the name of the higher culture, now by the name of Socialism, which has been voiced in our time by Ruskin and Morris, which has expressed itself in movements like the arts and crafts, or is revealed in the inspired paintings of the Pre-Raphaelites, demands just such a condition as in India our commercialization is destroying. The spiritual reawakening in the West is appealing for a social condition in which each man shall have not only an economic but a spiritual status in the society in which he lives, or as some of us would prefer to put it, he shall have a stable economic status in order that he may have a spiritual status as well.

It is such a condition that still exists in India, where society is organized upon a basis of "personal responsibility and cooperation" instead of, as with us, upon a basis of contract and competition. Even if we admit that the change in the Aryan of the West from the one basis to the other has been necessary in order to produce the conditions of modern progress, the scientific results of our civilization, in short, the great ships, it may yet be that the spiritual reawakening that is beginning to stir the dry bones of our Western materialism may yet leave the ancient East fundamentally unchanged, and bring us once again into some kindred condition through our contact with her. . .

But has commerce any aim? Is it as yet more than a blind force? The experience of 30 years has shown that this appeal of the English artists might have been as eloquently made on behalf of English as of Indian craftsmanship. For, indeed, the appeal is less against English Governmental action than against the conditions imposed upon the world by the development of industrial machinery, directed by commercialism. Industrial machinery which is blindly displacing the purpose of hand and destroying the individuality of human production; and commercialism, which is setting up one standard only, the quantitative standard, the standard of saleability.

To compass the destruction of commercialism and regulate and delimit the province of industrial machinery for the benefit of mankind is now the work of the Western reformer. The spiritual re-awakening with us is taking that shape.

It is probable that in this effort of the Western artists, workmen and reformers for the reconstruction of society on a saner and more spiritual basis, the East can help us even more than we shall help the East. What would we not give in England for a little of that "workshop service" which Dr. Coomaraswamy describes, in place of our half-baked evening classes in County Council Schools? What, in our effort at the revival of handicrafts in the decaying countryside, for some of those "religious trade union villages" of which Sir George Birdwood speaks?

There has come over Western civilization, in the last 25 years, a green sickness, a disbelief, an unrest; it is not despondency, for in the finer minds it takes the form of an intense spiritual hopefulness; but it takes the form also of a profound disbelief in the value of the material conditions of modern progress, a longing to sort the wheat from the chaff, the serviceable from the useless, a desire to turn from mechanical industry and its wastefulness, and to look once more to the human hand, to be once again with Mother Earth.

"It behooves us," said Heraclitus, in the time of the beginnings of Hellenic civilization, "it behooves us to follow the common reason of the world; yet, though there is a common reason in the world, the majority live as though they possessed a wisdom peculiar each unto himself alone." This is so profoundly modern that it might almost be a comment upon English or American industrialism, did we not know that it applied equally to the peculiar intellectual individualism of Hellas, which disintegrated and destroyed her culture. But the "common reason of the world," if the words of Heraclitus are to be taken at their face value, includes the reason of

the East, and with it the social order that has stood there unshaken for 3,000 years, and hence stood there long before the days of Heraclitus himself.

For our immediate purpose, too, the purpose of this book, the "common reason of the world" includes and defines the Indian craftsman and the Indian village community; it gives them a definite and necessary place not only in the Indian order of things, not only in the culture of the East, but in the world. It shows them to be reasonable and right, and it shows them, what is still more important, to be the counterpart one of the other.

Here once more we are learning from the East. The English craftsman and the English village are passing, or have passed away; and it is only in quite recent times that we have discovered that they, too, are the counterpart one of the other. Industrial machinery, blindly misdirected, has destroyed them both, and recent English land legislation has been trying, with allotment and small-holding acts, to re-establish the broken village life. Those of us, however, that have studied the Arts and Crafts in their town and country conditions, are convinced that the Small Holding Problem is possible of solution only by some system of cooperation, and if some forms of the craftsmanship are simultaneously revived and added to it.

"Speak to the earth and it shall teach thee;" that is an old lesson, and it is true not only of England, but of all Western countries that have been touched by the greensickness of industrial machinery. ∎

When Tom Bender wrote Sharing Smaller Pies *in January, 1975, it was an important break-through from the woeful cries about resource shortages to a recognition of the lessons we had learned from our splurge. In its entirety, the paper is a look at the new possibilities and values inherent in this time of transition and beyond.* Sharing Smaller Pies *(7) is available from Rain (48). Reprinted with permission from the author.*

CHANGING POSSIBILITIES

By Tom Bender
From SHARING SMALLER PIES

PLENTIFUL RESOURCES have until recently freed us from having to judge—from the need for the wisdom or ethical strength to say, "This will not be good for us." Having the ability to mask the symptoms and effects of unwise actions, to rebuild structure cheaply, and to introduce exotic sources of materials and energy, has permitted us to try anything that held the promise of immediate benefit—or just be possible—regardless of its eventual cost or damage. This has had a positive aspect, for it has given us an opportunity to test our ethical and moral wisdom—no holds barred—and through our mistakes to come to a deeper and more fundamental basis for the choices we learn to be right.

It has allowed us to explore new kinds of technologies, social and political organization, and assumptions about our world. From testing our past assumptions, and through being able to repeat all the evolutionary dead ends that nature has already abandoned, we have the opportunity to come to a closer understanding of the real possibilities and limitations of our world. Such understanding can give us greater and more precisely defined freedom and more thorough and precise understanding of what we can and should do and not do. The closer we can move our arbitrary human laws to the reali-

ties of natural law, the less arbitrary and more meaningful they become, and less is required to enforce and sustain them. Within natural law is total freedom, for it defines the realities through which we must move.

Such a period of testing is always limited, and is coming to an end today as the inexpensive energy to support such experimentation is becoming less available, and as the externalized and deferred costs of the ways we have experimented with are becoming visible and unaffordable at increasingly rapid rates. This is forcing us as individuals and as a society to develop and choose values, to develop ways of living, working, relating and feeling which offer greatest benefit to our survival and well-being, and to develop the moral discipline to sustain such ways.

Our present assumptions reflect conditions of plenty and promises of continued and even greater plenty. Inexpensive energy has been available to multiply the effect of our work and make our dreams more easily achievable. Natural resources have been plentiful and easily obtained. Our population has been relatively small in comparison to these resources, and human resources thus in relatively short supply.

We have believed that human work could and should be replaced by machines, and that such changes would contribute to our well being. We have considered that enlargement of our material wealth offered the primary if not sole means to improve our quality of life—to the point of equating measurement of our material productive capabilities with quality of life.

The novelty and excitement of our new abilities to materially change the conditions of our lives through massive use of resources has made evaluation of such actions difficult. Inexperience with the effects of such profoundly different technologies has until recently made such evaluation impossible and reasonable control unattainable.

We have shown little concern about the economic and social effects of our actions upon each other and even less upon people of other cultures. We assumed that such problems would be resolved through greater production of goods and services, or were unimportant because of our relatively great material wealth and assumed equity of opportunity for everyone. We also have had great confidence that the apparently powerful tools of our sciences and technology could apply and have as positive an effect on other aspects of our lives as they have on our material well-being. We have felt that we could

beneficially institutionalize our individual responsibilities for caring for ourselves and others, for seeing to our health, education and safety, and for seeing to the effects of our actions.

Such assumptions have proven wrong.

We have plunged headlong for more than half a century into development of a technological direction unprecedented on this planet, and are now finding that its direction is unsustainable, its effects undesirable, and its replacement necessary if we are to ensure the soundness, stability and permanence of our society, our individual freedom and opportunity for personal growth and the overall health and well-being of our society. ∎

Should we feel guilty about our consumption of oil? Why feel guilty? If you do the simple solution is to stop using it. Let those who want to use it continue to do so as long as they can pay for it. We drilled for oil, we found it, we burned it, we have had a good time with it. What else can we expect of human beings? And, of course, the oil party won't last forever. Some people at the party are already feeling uncomfortable; most of the world would never even get near this extraordinary party.

If you feel uncomfortable with the oil, you can live a life that doesn't depend on it. Do this and know that others will follow as the oil runs out. But don't try to ruin the party! It will only make bad feelings and unhappiness. Let the Cadillacs rule while they can.

How can we live our lives without consuming huge quantities of oil? There are millions of Americans right now who don't use much oil, and many of them are actually happy people. Cutting back on oil consumption is frightening only to people who have no imagination and no confidence.

Steve Baer
From SILLY PRESIDENT CARTER, 1978 (3)

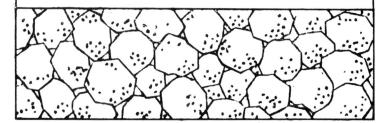

Stewart Brand knows how to sum things up. His Whole Earth Catalogs *(61) were watershed events responsible for sparking much of the interest in the issues that have become appropriate technology. They were also models for a brand new genre of information sharing. The pattern continues in* Co-Evolution Quarterly *(16) which reprinted many of the pieces in this book at one time or another.* Lessons *is published here with their permission from the Summer, 1975, issue.*

LESSONS

By Stewart Brand

THANKS TO THE TRIPLE BLESSING of the lost cause in Vietnam, the loss of control in the Energy Crisis, and the loss of innocence with Watergate, general prospects for America are now better than they've been in years. Three public humiliations in two years has put the country in Learning Mode.

The lesson of Watergate is Never Trust a Government. Already we are seeing improvement in the quality of representative that a burned, skeptical, informed electorate can select. Unfortunately, this process has also made us obsessed with government and the media. There's a problem box and a solution box, Ken Kesey maintains. No matter how earnestly you look, you won't find your solutions in the problem box.

The lesson of the Energy Crisis is If You Spend Capital (oil) As If It Were Income, You Live a Short Flashy Life. America is third-generation rich, famously wastrel. The Arabs are first-generation rich, with riches we had considered ours. In the resulting delicious comeuppance we have been forced to reassess where "ours" begins and ends. With the limits to material growth in sight, a central madness—cornucopia—begins to get well.

The lesson of Vietnam has to do with money versus turf, or even with money versus virtue. We

had grown used to buying our victories. In the U.S. Civil War the North sent more shoes and cannon to the South than the Confederate horsemen could keep up with—and a disastrous victory resulted. Later in the Atlantic, the amazing German U-boats were simply flooded with klutzy Liberty Ship targets—and we victors went crazy in a world of our own (apparent) making. The world was purchasable, and we had the cash. Vietnam proved otherwise.

So it's back to basics. The quality of land—sustained rather than spent. Homeland. The quality of people—their education, responsibility, humor. The blend of dependence and independence among citizens and among nations that keeps open the choices between cooperation, competition and count-me-out.

China has made national count-me-out look very effective indeed.

An old joke says that the lake with the longest name in the world is called—in a single native American word—"You-fish-on-your-side-we'll-fish-on-our-side-nobody-fish-in-the-middle."

"Nobody fish in the middle" is a formula for a perpetually livable planet.   ∎

11

12

NET ENERGY

When energy was cheap and abundant, it didn't matter how much was consumed in the process of using appliances, heating homes or making cars. There seemed to be plenty of the juicy stuff we were using to go around. But when supplies were cut during the Oil Embargo in 1973 and prices rose drastically, it became more important to look at the quality of energy and how efficiently it was being used for what purposes. One of the ways of doing this is with a tool called net energy analysis.

The concept of net energy, as developed by Dr. Howard Odum at the University of Florida, brought with it a flash of understanding to people who knew intuitively that something was wrong with those cheerful figures which reported that coal and oil would last well into the next century even though our growth rate was moving off the charts. The idea behind net energy is simple—it takes energy to get energy. It takes energy to strip every mine, drill every well and build every derrick. Only after the energy costs have been subtracted from the total can you tell how much energy is really available.

Net energy analysis is as much a way of judging when we are spending more than we are getting back as it is a way of understanding the relationships between various parts of a system and how they affect each other. It is also a means of determining the value of energy and how carefully it is being used. Wilson Clark's article, It Takes Energy to Get Energy—The Law of Diminishing Returns Is in Effect (reprinted here), is still one of the clearest translations of Odum's ideas about net energy.

Long before the Oil Embargo, Odum was exploring the relationships between energy, the health of ecosystems and the stability of the world economy. Energy, Ecology and Economics was the first presentation of Odum's ideas in a shorter, less technical form. Written about the time of the Embargo in 1973, it explores the political implications of our steadily increasing dependence on finite fossil fuel resources. It quickly became one of those much-reproduced papers that provided a theoretical framework for the search for innovative energy alternatives—by then already well underway.

In this paper, Odum takes a dim view of efforts to use mechanical devices to trap and use the energy from the sun and wind. His point is that such developments are bad imitations of biological processes that will never get off the ground without heavy fossil fuel subsidies. Yet, that begs the question of what we are to do as the well runs dry. Ironically, the arguments in the rest of the paper have been used by others to make the case that it is far wiser to use our remaining energy supplies to make the transition to a solar-based system than to build a nuclear future.

Cosmic Economics and Independence? were written by Joel Schatz and Tom Bender in 1974, shortly after Odum's Energy, Ecology and Economics appeared. They were working at the Oregon Office of Energy Research and Planning under former Governor Tom McCall to apply the concepts of net energy to state government planning. They picked up on Odum's hint that rising energy prices (due to decreasing net energy) are part of what fuels economic inflation. They put it as succinctly as possible to give immediate direction to efforts to cope with the impacts (both present and long-range) of the Oil Embargo. Looking back, it is interesting to note how the validity of these ideas has been confirmed. If this piece is correct, it is possible to save energy, save money and have jobs at the same time. The implications have yet to be spelled out with all the statistical data and dollar signs to convince the sceptics, but if you read between the lines, you'll find some eye-opening ideas. □

In the early 1970s, Smithsonian Magazine *did a series of articles on energy and appropriate technology written by Wilson Clark. This one on net energy appeared in the December 1974 issue with illustrations by Jan Adkins. Clark's book,* Energy for Survival *(14), has been called encyclopedic, which is no exaggeration. It is the best place to find information about the history, net supplies, use and value of various forms of energy. Reprint with permission from* Smithsonian Magazine *(55).*

IT TAKES ENERGY TO GET ENERGY: THE LAW OF DIMINISHING RETURNS IS IN EFFECT

By Wilson Clark

IN THE MID-NINETEENTH CENTURY a British company launched the *Great Eastern*, a coal-fired steamship designed to show the prowess of Britain's industrial might. The ship, weighing 19,000 tons and equipped with bunkers capable of holding 12,000 tons of coal, was to voyage to Australia and back without refueling. But it was soon discovered that to make the trip the ship would require 75 percent more coal than her coal-storage capacity—more coal, in fact, than the weight of the ship herself.

Today the United States is embarking on an effort to become independent in energy production, and such a program deserves the kind of analysis that the British shipbuilders overlooked. Indeed, our civilization appears to have reached a limit similar to that of the *Great Eastern*: The energy which for so long has driven our economy and altered our way of life is becoming scarce, and a number of respected experts are suggesting that, without significant changes, our society will go the way of the ship that needed more fuel than it could carry.

In recent years, energy growth in the United States has expanded at a rate of nearly four percent per year, resulting in a per capita consumption of all forms of energy higher than that of any other nation. U.S. energy consumption in 1970 was half again as much as all of Western Europe's, even though Europe's population is one-and-a-half times ours.

As energy consumption has increased in this nation, our energy resources have drastically declined. According to M. King Hubbert, a highly respected energy and resource expert, the peak for production of all kinds of liquid fossil fuel resources (oil and natural gas) was reached in this country in 1970 when almost four billion barrels were produced. "The estimated time required to produce the middle 80 percent [of the known reserves of this resource]," Hubbert says, "is the 61-year period from 1939 to the year 2000, well under a human lifespan."

As available domestic oil and gas resources have declined, we have turned more and more to foreign imports—but, since 1973, the price of this essential imported oil has quadrupled. . .

The central problem is simply that *it takes energy to produce new energy*. In other words, in every process of energy conversion on Earth, some energy is inevitably wasted. The laws of thermodynamics, formulated in the last century, might be viewed as

14

describing a sort of "energy gravity" in the universe: energy constantly moves from hot to cold, from a higher to a lower level. Some energy is free for Man's use—but it must be of high quality. Once used, it cannot be recycled to produce more power.

Coal, for example, can be burned in a power plant to produce steam for conversion into electric power. But the resulting ashes and waste heat cannot be collected and burned to produce yet more electricity. The quality of the energy in the ashes and heat is not high enough for further such use.

Numerous studies have indicated that the United States has enormous reserves of fossil fuels which can provide centuries of energy for an expanding economy, yet few take into account the thermodynamic limitations on mining the fuels left. Most cheap and accessible fuel deposits have already been exploited, and the energy required to fully exploit the rest may be equal to the energy contained in them. What is significant, and vital to our future, is the *net* energy of our fuel resources, not the *gross* energy. Net energy is what is left after the processing, concentrating and transporting of energy to consumers is subtracted from the gross energy of the resources in the ground.

Consider the drilling of oil wells. America's first oil well was drilled in Pennsylvania in 1859. From 1860 to 1870, the average depth at which oil was found was 300 feet. By 1900, the average find was at 1,000 feet. By 1927, it was 3,000 feet; today, it is 6,000 feet. Drilling deeper and deeper into the earth to find scattered oil deposits requires more and more energy. Think of the energy costs involved in building the trans-Alaska pipeline. For natural gas, the story is similar.

Dr. Earl Cook, dean of the College of Geosciences at Texas A. & M. University, points out that drilling a natural gas well doubles in cost each 3,600 feet. Until 1970, he says, all the natural gas found in Texas was no more than 10,000 feet underground, yet today the gas reserves are found at depths averaging 20,000 feet and deeper. Drilling a typical well less than a decade ago cost $100,000, but now the deeper wells each cost more than $1,000,000 to drill. As oilmen move offshore and across the globe in their search for dwindling deposits of fossil fuels, financial costs increase, as do the basic energy costs of seeking the less concentrated fuel sources.

Although there is a good deal of oil and natural gas in the ground, the net energy—our share—is decreasing constantly.

The United States has deposits of coal estimated at 3.2 trillion tons, of which up to 400 billion tons may be recoverable—enough, some say, to supply this nation with coal for more than 1,000 years at present rates of energy consumption. And since we are dependent on energy in liquid and gaseous form (for such work as transportation, home and industrial heating), the energy industries and the Federal Energy Administration* have proposed that our vast coal deposits be mined and then converted into gas and liquid fuels.

Yet the conversion of coal into other forms of energy, such as synthetic natural gas, requires not only energy but large quantities of water. In fact, a panel of the National Academy of Sciences recently reported that a critical water shortage exists in the Western states, where extensive coal deposits are located. "Although we conclude that enough water is available for mining and rehabilitation at most sites," said the scientists, "not enough water exists for large-scale conversion of coal to other energy forms (*e.g.*, gasification or steam electric power). The potential environmental and social impacts of the use of this water for large-scale energy conversion projects would exceed by far the anticipated impact of mining alone." In fact, the energy and water limitations in the Western states preclude more than a fraction of the seemingly great U.S. coal deposits from ever being put to use for gasification or liquefaction.

The prospects for oil shale development are not as optimistic as some official predictions portend. Unlike oil, which can be pumped from the ground relatively easily and refined into useful products, oil shale is a sedimentary rock which contains kerogen, a solid, tarlike organic material. Shale rock must be mined and heated in order to release oil from kerogen. The process of mining, heating and processing the oil shale requires so much energy that many experts believe that the net energy yield from shale will be negligible. According to *Business Week*, at least one major oil company has decided that the net energy yield from oil shale is so small that they will refuse to bid on federal lands containing deposits. And even if a major oil-shale industry were to develop, water supplies would be as great a problem as for coal conversion, since the deposits are in water-starved Western regions. The twin limiting factors of water and energy will preclude the substantial development of these industries.

---

*ed. note: now the Federal Department of Energy.

Nuclear power is seen as the key to the future, yet an energy assessment of the nuclear fuel cycle indicates that the net energy from nuclear power may be more limited than the theoretically prodigious energy of the atom has promised.

Conventional nuclear fission power plants, which are fueled by uranium, contribute little more than four percent of the U.S. electricity requirements at present, but according to the Atomic Energy Commission, fission will provide more than half of the nation's electricity by the end of the century. Several limitations may prevent this from occurring. One is the availability of uranium ore in this country for conversion to nuclear fuel. According to the U.S. Geological Survey, recoverable uranium resources amount to about 273,000 tons, which will supply the nuclear industry only up to the early 1980s. After that, we may well find ourselves bargaining for foreign uranium, much as we bargain for foreign oil today.

According to energy consultant E. J. Hoffman, however, an even greater problem with nuclear power is that the fuel production process is highly energy-intensive. "When all energy inputs are considered," he says, such as mining uranium ore, enriching nuclear fuel, and fabricating and operating power plants and reprocessing facilities, "the net electrical yield from fission is very low." Optimistic estimates from such sources as the President's Council on Environmental Quality say that nuclear fission yields about 12 percent of the energy value of the fuel as electricity: Hoffman's estimate is that it yields only 3 percent. That advanced reactors might have a higher net yield is one potential, but largely unknown at present, since such reactors have not yet been built and operated commercially. Other nuclear power processes, such as nuclear fusion, have simply not yet been shown to produce electricity, and so they cannot be counted upon. Even the more "natural" alternative energy sources, such as solar power, wind power and geothermal power,

have not been evaluated from the net energy standpoint. They hold out great promise—especially from a localized, small-scale standpoint. Solar energy, for example, is enormous on a global scale, but its effect varies from one place to another. However, the net energy yield from solar power overall might be low, requiring much energy to build elaborate concentrators and heat storage devices necessary.

What about hydrogen as a replacement fuel? By itself, hydrogen is not at all abundant in nature, and other energy sources must first be developed to power electrolyzers in order to break down water into hydrogen and oxygen. The energy losses inherent in such processes may result in a negligible overall energy yield by the time hydrogen is captured, stored and then burned as fuel. An indication of the magnitude of this problem has been given by Dr. Derek Gregory of the Institute of Gas Technology in Chicago, who points out that to substitute hydrogen fuel fully for the natural gas currently produced would require the construction of 1,000 enormous one-million-kilowatt capacity electric power plants to power electrolyzers—more than twice the present entire installed electrical plant capacity of the nation.

While much of this kind of analysis is apparently new to most energy planners, it also represents more than an analogy to the cost-accounting that is familiar to businessmen investing dollars to achieve a net profit. The net energy approach might provide a new way of looking at subjects so seemingly disparate as the natural world and the economy.

DOLLAR VALUES OF NATURAL SYSTEMS

An outspoken proponent of the net energy approach is Dr. Howard T. Odum, a systems ecologist at the University of Florida. In the 1950s, Odum analyzed the work of researchers trying to grow algae as a cheap source of fuel, and found that the energy required to build elaborate facilities and maintain algae cultures was greater than the energy yield of the algae when harvested for dry organic material. The laboratory experiment was subsidized, not by algae feeding on free solar energy—which might have yielded a net energy return—but by "the fossil fuel culture through hundreds of dollars spent annually on laboratory equipment and services to keep a small number of algae in net yields."

With his associates at the University of Florida, Odum began to develop a symbolic energy language, using computer-modeling techniques, which relates energy flows in the natural environment to the energy flows of human technology.

Odum points out that natural sources of energy —solar radiation, the winds, flowing water and energy stored in plants and trees—have been treated as free "gifts" rather than physical energy resources which we can incorporate into our economic and environmental thinking. In his energy language, however, a dollar value is placed on all sources of energy—whether from the sun or petroleum. To produce each dollar in the economy requires energy—for example, to power industries. The buying power of the dollar, therefore, can be given an energy value. On the average, Odum cal-

culates, the dollar is worth 25,000 calories (kilocalories, or large calories) of energy—the familiar energy equivalent dieters know well as food values. Of this figure, 17,000 calories are high-quality energy from fossil fuels and 8,000 calories low-quality energy from "natural" sources. In other words, the dollar will buy work equal to some mechanical labor, represented by fossil fuel calories, and work done by natural systems and solar energy.

Odum's concept of energy as the basis of money is not new; a number of 19th-century economists thought of money or wealth as deriving from energy in nature. The philosophy was expounded earlier in this century by Sir Frederick Soddy, the British scientist and Nobel Laureate, who wrote that energy was the basis of wealth. "Men in the economic sense," he said, "exist solely by virtue of being able to draw on the energy of nature. . . . Wealth, in the economic sense of the physical requisites that enable and empower life, is still quite as much as of yore the product of the expenditure of energy or work."

Odum views natural systems as valuable converters and storage devices for the solar energy which triggers the life-creating process of photosynthesis. Even trees can be given a monetary value for the work they perform, such as air purification, prevention of soil erosion, cooling properties, holding ground water, and so on. In certain locations, he says, an acre of trees left in the natural state is worth more than $10,000 per year or more than $1 million over a hundred-year period, not counting inflation.

Last year, he calculated that solar energy, in con-

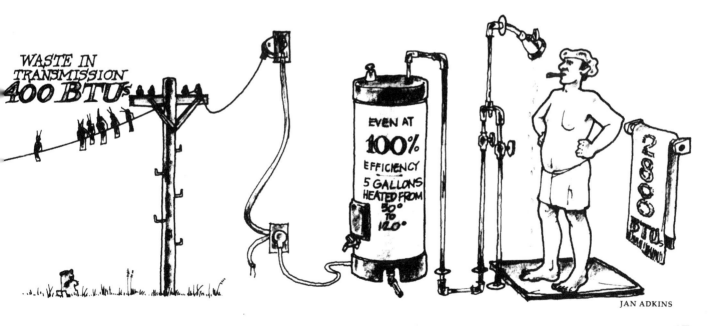

JAN ADKINS

17

junction with winds, tides and natural ecological systems in the state of Florida, contributed a value of $3 billion to the state, compared to fossil fuel purchases by the state's citizens of $18 billion per year.

The value of the natural systems to the state had never before been calculated. "These parts of the basis of our life," says Odum, "continue year after year, diminished, however, when ecological lands that receive sun, winds, waves and rain are diverted to other use.". . .

Odum's work may lead to eminently practical applications, by indicating directions in which our society can make the best use of energy sources and environmental planning. One application is to use natural systems for treating wastes, rather than using fossil fuels to run conventional waste-treatment plants. "There are," he says, "ecosystems capable of using and recycling wastes as a partner of the city without drain on the scarce fossil fuels. Soils take up carbon monoxide, forests absorb nutrients, swamps accept and regulate floodwaters." He is currently involved in a three-year program in southern Florida to test the capability of swamps to treat wastes, and demonstrate their value to human civilization as a natural "power plant." The work, supported by the Rockefeller Foundation and the National Science Foundation, has drawn the attention and interest of many community and state governments.

According to Odum's energy concepts, a primary cause of inflation in this country and others is the pursuit of high economic growth with ever-more costly fossil fuels and other energy sources. As we dig deeper in our search for less-concentrated energy supplies to fuel our economy, the actual value of our currency is lessening. "Because so much energy has to go immediately into the energy-getting process," he notes, "then the real work to society per unit of money is less."

Economists, who generally resent intruders on their turf, have not embraced this equation of energy and money with much enthusiasm, but it is gaining adherents in several quarters.

• • •

Since the Industrial Revolution, the Western world has been engaged in a great enterprise—the building of a highly complicated technological civilization. The Western "growth" economy (which today also characterizes Japan) has been made possible by seemingly endless supplies of inexpensive energy. One implication of the net energy approach is that a vigorous and wide-reaching conservation program may be the only palliative for inflation.

Another implication is that the days of high growth may be over sooner than most observers have previously thought. For it is increasingly apparent that today's energy crisis is pushing us toward a "steady-state" economy: No one yet knows what such an economy will look like or what social changes will result. But it would seem to be about time to start thinking seriously about it. ∎

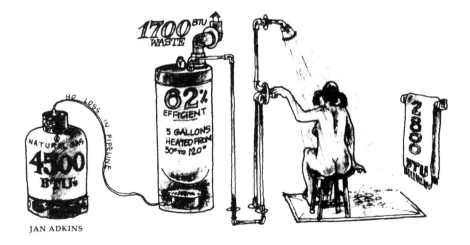

JAN ADKINS

This important piece was originally presented as a speech to the Royal Swedish Academy of Science in August, 1974, and was published in their journal, Ambio *(1), in the fall of that year. For more detailed information on Odum's work on net energy see* Environment, Power and Society *(45) or his more recent book,* Energy Basis for Man and Nature *(46), written with his wife, Elizabeth. Reprinted with permission from* Ambio.

ENERGY, ECOLOGY & ECONOMICS

By Howard T. Odum

As LONG PREDICTED ENERGY SHORTAGES appear, as questions about the interaction of energy and environment are raised in legislatures and parliaments, and as energy-related inflation dominates public concern, many are beginning to see that there is a unity of the single system of energy, ecology and economics. The world's leadership, however, is mainly advised by specialists, those who study only a part of the system at a time, and instead of a single systems understanding, we have adversary arguments and special pleadings mainly fallacious and dangerous to the welfare of nations and the role of man as the earth's information bearer and programmatic custodian. Many economic models ignore the changing force of energy because energy sources have always been regarded as an external constant; ecoactivists cause governments to waste energy in unnecessary technology; and the false gods of growth and medical ethics make famine, disease and catalytic collapse more and more likely for much of the world. Some energy specialists consider the environment as an antagonist instead of a major energy ally in supporting the biosphere.

Here are the main points that we must get across to our leadership:

1. The true value of energy to society is the net energy, which is that after the energy costs of getting and concentrating that energy are subtracted. Many forms of energy are low-grade because they have to be concentrated, transported, dug from deep in the earth or pumped from far at sea. Much energy has to be used directly and indirectly to support the machinery, people, supply systems, etc. to deliver the energy. If it takes 10 units of energy to bring 10 units of energy to the point of use, then there is no net energy. Right now we dig further and further, deeper and deeper, and go for energies that are more and more dilute in the rocks. We are still expanding our rate of consumption of gross energy, but since we are feeding a higher and higher percentage back into the energy-seeking process, we are decreasing our percentage of net energy production. Many of our proposed alternative energy sources take more energy feedback than present processes.

19

2. Worldwide inflation is driven in part by the increasing fraction of our fossil fuels that have to be used in getting more fossil and other fuels. If the money circulating is the same or increasing, and if the *quality energy* reaching society for its general work is less because so much energy has to go immediately into the energy getting process, then the real work to society per unit of money circulated is less. Money buys less real work of other types and thus money is worth less. Because the economy and total energy utilization are still expanding, we are misled to think the total value is expanding and we allow more money to circulate, which makes the money-to-work ratio even larger.

3. Many calculations of energy reserves which are supposed to offer years of supply are as gross energy rather than net energy and thus may be of much shorter duration than often stated. Suppose for every 10 units of some quality of oil shale proposed as an energy source there were required 9 units of energy to mine, process, concentrate, transport, and meet environmental requirements. Such a reserve would deliver 1/10 as much net energy and last 1/10 as long as was calculated. Leaders should demand of our estimators of energy reserves that they make their energy calculations in units of net energy. Many calculations are made in terms of reserves available for a given price, which is a step in the right direction, but since the prices are also inflated by the energy shortages, such calculations compound the errors. Net energy should be calculated as gross energy minus the energy cost of concentrating and delivering the energy. The net reserves of fossil fuels are mainly unknown, but they are much smaller than the gross reserves which have been the basis of public discussions and decisions that imply that growth can continue. Much of the so-called energy reserves may not even be available because the energy cost of collecting, concentrating and bringing energy to use sites is greater than the energy remaining to be used.

4. Societies compete for economic survival by Lotka's principle, which says that systems win and dominate that maximize their useful total power from all sources and flexibly distribute this power toward needs affecting survival. The programs of

forests, seas, cities and countries survive that maximize their system's power for useful purposes. The first requirement is that opportunities to gain inflowing power be maximized, and the second requirement is that energy utilization be effective and not wasteful as compared to competitors or alternatives.

5. During times when there are opportunities to expand one's power inflow, the survival premium by Lotka's principle is on rapid growth even though there may be waste. We observe dog-eat-dog growth competition every time new vegetation colonizes a bare field where the immediate survival premium is first placed on rapid expansion to cover the available energy receiving surfaces. The early growth eco-systems put out weeds of poor structure and quality, which are wasteful in their energy-capturing efficiencies, but effective in getting growth even though the structures are not long lasting. Most recently modern communities of man have been in 200 years of colonizing growth, expanding to new energy sources such as fossil fuels, agricultural lands and other special energy sources. Western culture, and more recently Eastern and Third World cultures, are locked into a mode of belief in growth as necessary to survival. "Grow or perish" is what Lotka's principle requires, *but* only during periods when there are energy sources which are not yet tapped.

6. During times when energy flows have been tapped and there are no new sources, Lotka's principle requires that those systems win that do not allow fruitless growth but instead use all available energies in long-staying, high diversity, steady state works. Whenever an eco-system reaches its steady state after periods of succession, the rapid net growth specialists are replaced by a new team of higher diversity, higher quality, longer living, better controlled and stable components. Collectively through division of labor and specialization, the climax team gets more energy out of the steady flow of available source energy than those specialized in fast growth could.

Our system of man and nature will soon be shifting from a rapid growth criterion of economic survival to steady state non-growth as the criterion for maximizing one's work for economic survival. The timing depends only on the reality of one or two possibly high-yielding nuclear energy processes (fusion and breeder reactions) which may not be very yielding.

Ecologists are familiar with both growth states and steady state and observe both in natural systems in their work routinely, but economists were all trained in their subject during rapid growth, and most don't even know there is such a thing as steady state. Most economic advisors have never seen a steady state even though most of man's million-year history was close to steady state. Only the last two centuries have seen a burst of temporary growth because of temporary use of special energy supplies that accumulated over long periods of geologic time.

7. High quality of life for humans and equitable economic distribution are more closely approximated in steady state than in growth periods. During growth, emphasis is on competition, and large differences in economic and energetic welfare develop; competitive exclusion, instability, poverty and unequal wealth are characteristic. During steady state, competition is controlled and eliminated, being replaced with regulatory systems, high division and diversity of labor, uniform energy distributions, little change, and growth only for replacement purposes. Love of stable system quality replaces love of net gain. Religious ethics adopt something closer to that of those primitive peoples that were formerly dominant in zones of the world with cultures based on the steady energy flows from the sun. Socialistic ideals about distribution are more consistent with steady state than growth.

8. The successfully competing economy must use its net output of richer quality energy flows to subsidize the poorer quality energy flow so that the total power is maximized. In ecosystems, diversity of species develop that allow more of the energies to be tapped. Many of the species that are specialists in getting lesser and residual energies receive subsidies from the richer components. For example, the sunny leaves on tops of trees transport fuels that help the shaded leaves so they can get some additional energy from the last rays of dim light reaching the forest floor. The system that uses its excess energies in getting a little more energy, even from sources that would not be net yielding alone, develops more total work and more resources for total survival. In similar ways we now use our rich fossil fuels to keep all kinds of goods and services of our economy cheap so that the marginal kinds of energies may receive the subsidy benefit that makes them yielders, whereas they would not be able to generate much without the subsidy.

9. Energy sources which are now marginal, being supported by hidden subsidies based on fossil fuel, become less economic when the hidden subsidy is removed. A corollary of the previous principle of using rich energies to subsidize marginal ones is that the marginal energy sources will not be as net yielding later since there will be no subsidy. This truth is often stated backwards in economists' concepts because there is inadequate recognition of external changes in energy quality. Often they propose that marginal energy sources will be economic later when the rich sources are gone. An energy source is not a source unless it is contributing to net yields, and ability of marginal sources to yield goes down as the other sources of subsidy become poorer.

10. Increasing energy efficiency with new technology is not an energy solution since most technological innovations are really diversions of cheap energy into hidden subsidies in the form of fancy, energy-expensive structures. Most of our century of progress with increasing efficiencies of engines has really been spent developing mechanisms to subsidize a process with a second energy source. Many calculations of efficiency omit these energy inputs. We build better engines by putting more energy into the complex factories for manufacturing the equipment. The percentage of energy yield in terms of all the energy incoming may be less, not greater. Making energy net yielding is the only process not amendable to high energy-based technology.

11. Even in urban areas more than half of the useful work on which our society is based comes from the natural flows of sun, wind, waters, waves, etc. that act through the broad areas of seas and landscapes without money payment. An economy to compete and survive must maximize its use of these energies, not destroying their enormous free subsidies. Necessity of environmental inputs are often not realized until they are displaced. When an area first grows, it may add some new energy sources in fuels and electric power, but when it gets to about 50% of the area developed it begins to destroy and diminish as much necessary life support work that was free and unnoticed as it adds. At this point, further growth may produce a poor ability in economic competition because the area now has higher energy drains. For example, areas that grow too dense with urban developments may pave over the areas that formerly accepted and reprocessed waste waters. As a consequence, special tertiary waste treatments become necessary and monetary and energy drains are diverted from useful works to works that were formerly supplied free.

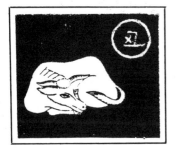

12. Environmental technology which duplicates the work available from the ecological sector is an economic handicap. As growth of urban areas has become concentrated, much of our energies and research and development work has been going into developing energy-costing technology to protect the environment from wastes, whereas most wastes are themselves rich energy sources for which there are, in most cases, ecosystems capable of using or recycling wastes as a partner of the city without drain on the scarce fossil fuels. Soils take up carbon monoxide, forests absorb nutrients, swamps accept and regulate floodwaters. If growth is so dense that environmental technology is required, then it is too dense to be economically vital for the combined system of man and nature there. The growth needs to be arrested or it will arrest itself with depressed economy of man and his environs. For example, there is rarely excuse for tertiary treatment because there is no excuse for such dense packing of growth that the natural buffer lands cannot be a good cheap recycling partner. Man as a partner of nature must use nature well, and this does not mean crowd it out and pave it over; nor does it mean developing industries that compete with nature for the waters and wastes that would be an energy contributor to the survival of both.

13. Solar energy is very dilute, and the inherent energy cost of concentrating solar energy into form for human use has already been maximized by forests and food producing plants. There is no yield from the sun possible beyond the familiar yields from forestry and agriculture. Advocates of major new energies available from the sun don't understand that the concentration quality of solar energy is very low, being only 10^{-16} kilocalories per cubic centimeter. Plants build tiny microscopic semiconductor photon receptors that are the same in principle as the solar cells advocated at vastly greater expense by some solar advocates. The reason solar technology has not and will not be a major contributor or substitute for fossil fuels is that it will not compete without energy subsidy from fossil fuel economy. The plants have already maximized use of sunlight. Higher yields require large fossil fuel subsidies in making the solar receiving structures, whereas the plants make their equipment out of the energy budget they process. Since man has already learned how to subsidize agriculture and forestry with fossil fuels when he has it, solar technology becomes a duplication. The reason major solar technology has not and will not be a major contributor or substitute for fossil fuels is that it will not compete without energy subsidy from the fossil fuel economy. Some energy savings are possible in house heating on a minor scale.

14. Energy is measured by calories, BTUs, kilowatt hours and other introconvertible units, but energy has a scale of quality which is not indicated by these measures. The ability to do work for man depends on the energy quality and quantity, and this is measurable by the amount of energy of a lower quality grade required to develop the higher grade. The scale of energy goes from dilute sunlight up to plant matter to coal and oil to electricity and up to the high quality effort of computer and human information processing.

15. Nuclear energy is now mainly subsidized with fossil fuels and barely yields net energy. High costs of mining, processing fuels, developing costly plants, storing wastes, operating complex safety systems and operating government agencies make present nuclear energy one of the marginal sources which add some energy now, while they are subsidized by a rich economy. A self-contained isolated nuclear energy does not now exist. Since the present nuclear energy is marginal while it uses the cream of rich fuels accumulated during times of rich fossil fuel excess, and because the present rich reserves of nuclear fuel will last no longer than fossil fuels, there may not be major long-range effect of present nuclear technology on economic survival. High energy cost of nuclear construction may be a factor accelerating the exhaustion of the richer fuels.

BREEDER PROCESS: The Breeder Process is now being given its first tests of economic effectiveness and we don't yet know how *net* yielding it will be. The present nuclear plants are using up the fuels that could support the breeder reactors if these turn out to be net yielders over and beyond the expected high energy costs in safety costs, occasional accidents, reprocessing plants, etc. Should we use the last of our rich fossil fuel wealth for the high re-

search and development costs and high capital investments of processes too late to develop a net yield?

FUSION: The big question is will fusion be a major net yield? The feasibility of pilot plants with the fusion process is unknown. There is no knowledge yet as to the net energy in fusion or the amounts of energy subsidy fusion may require. Because of this uncertainty, we cannot be sure about the otherwise sure—leveling and decline in total energy flows that may soon be the pattern for our world.

16. Substantial energy storages are required for stability of an economy against fluctuations of economies, or of natural causes, and of military threats. The frantic rush to use the last of the rich oils and gas that are easy to harvest for a little more growth and tourism is not the way to maintain power stability or political and military security for the world community of nations as a whole. World stability requires a deenergizing of capabilities of vast war and an evenly distributed power basis for regular defense establishments, which need to be evenly balanced without great power gradients that encourage change of military boundaries. A two-year storage is required for stability.

17. The total tendency for net favorable balance of payments of a country relative to others depends on the relative net energy of that country including its natural and fuel-based energies minus its wastes and non-productive energy uses. Countries with their own rich energies can export goods and services with less requirement for money than those that have to use their money to buy their fuels.

23

Those countries with inferior energy flows into useful work become subordinate energy dependents to other countries. A country that sells oil but does not use it within its boundaries to develop useful work is equally subordinate since a major flow of necessary high quality energy in the form of technical goods and services is external in this case. The country with strongest position is the one with a combination of internal sources of rich energies and internal sources of developed structure and information based on the energy.

18. During periods of expanding energy availabilities, many kinds of growth-priming activities may favor economic vitality and the economy's ability to compete. Institutions, customs and economic policies aid by accelerating energy consumption in an autocatalytic way. Many pump priming properties of fast-growing economies have been naturally selected and remain in procedures of government and culture. Urban concentrations, high use of cars, economic subsidy to growth, oil depletion allowances, subsidies to population growth, advertising, high-rise building, etc., are costly in energy for their operation and maintenance, but favor economic vitality as long as their role as pump primers is successful in increasing the flow of energy over and beyond their special cost. Intensely concentrated densities of power use have been economic in the past because their activities have accelerated the system's growth during a period when there were new energy sources to encompass.

19. During periods when expansion of energy sources is not possible, then the many high density and growth-promoting policies and structures be- come an energy liability because their high energy cost is no longer accelerating energy yield. The pattern of urban concentration and the policies of economic growth simulation that were necessary and successful in energy growth competition periods are soon to shift. There will be a premium against the use of pump priming characteristics since there will be no more unpumped energy to prime. What did work before will no longer work, and the opposite becomes the pattern that is economically successful. All this makes sense and is commonplace to those who study various kinds of ecosystems, but the economic advisors will be sorely pressed and lose some confidence until they learn about the steady state and its criteria for economic success. Countries with great costly investments in concentrated economic activity, excessive transportation customs, and subsidies to industrial expansion will have severe stresses. Even now the countries who have not gone so far in rapid successional growth are setting out to do so at the very time when their former more steady state culture is about to begin to become a more favored economic state comparatively.

20. Systems in nature are known that shift from fast growth to steady state gradually with programmatic substitution, but other instances are known in which the shift is marked by total crash and destruction of the growth system before the emergence of the succeeding steady state regime. Because energies and monies for research, development and thinking are abundant during growth and not during energy leveling or decline, there is a great danger that means for developing the steady state will not be ready when they are needed, which may be no more than five years away, but probably more like twenty years. (If fusion energy is a large net energy yielder, there may be a later growth period when the intensity of human power development begins to affect and reduce the main life support systems of the oceans, atmospheres and general biosphere.)

The humanitarian customs of the earth's countries now in regard to medical aid, famine and epidemic are such that no country is allowed to develop major food and other critical energy shortage because the others rush in their reserves. This practice has insured that no country will starve in a major way until we all starve together when the reserves are no longer available.

Chronic disease has evolved with man as his regulator, being normally a device for infant mortality and merciful old age death. It provided on the average an impersonal and accurate energy testing of body vitalities, adjusting the survival rate to the energy resources. Even in the modern period of high energy medical miracles, the energy for total medical care systems is a function of the total country's energies, and as energies per capita fall again so will the energy for medicine per capita, and the role of disease will again develop its larger role in the population regulation system. Chronic disease at its best was and is a very energy-inexpensive regulator.

Epidemic disease is something else. Nature's systems normally use the principle of diversity to eliminate epidemics. Vice versa, epidemic disease is nature's device to eliminate monoculture, which may be inherently unstable. Man is presently allowed the special high yields of various monocultures including his own high density population, his paper source in pine trees, and his miracle rice only so long as he has special energies to protect these artificial ways and substitute them for the disease which would restore the high diversity system, ultimately the more stable flow of energy.

The terrible possibility that is before us is that there will be the continued insistence on growth with our last energies by the economic advisors that don't understand, so that there are no reserves with which to make a change, to hold order, and to cushion a period when populations must drop. Disease reduction of man and of his plant production systems could be planetary and sudden if the ratio of population to food and medical systems is pushed to the maximum at a time of falling net energy. At some point the great gaunt towers of nuclear energy installations, oil drilling and urban cluster will stand empty in the wind for lack of enough fuel technology to keep them running. A new cycle of dinosaurs will have passed its way. Man will survive as he reprograms readily to that which the ecosystem needs of him so long as he does not forget who is serving whom. What is done well for the ecosystem is good for man. However, the cultures that say only what is good for man is good for nature will pass and be forgotten like the rest.

There was a famous theory in paleoecology called orthogenesis, which suggested that some of the great animals of the past were part of systems that were locked into evolutionary mechanisms by

which the larger ones took over from smaller ones. The mechanisms then became so fixed that they carried the size trend beyond the point of survival, whereupon the species went extinct. Perhaps this is the main question of ecology, economics and energy. Has the human system frozen its direction into orthogenetic paths to cultural crash, or is the great creative activity of the current energy-rich world already sensing the need for change with the alternatives already being tested by our youth so they will be ready for the gradual transition to a fine steady state that carries the best of our recent cultural evolution into new, more miniaturized, more dilute and more delicate ways of man and nature?

In looking ahead, the United States and some other countries may be lucky to be forced by changing energy availabilities to examine themselves, level their growth, and change their culture towards the steady state early enough so as to be ready with some tested designs before the world as a whole is forced to this. A most fearful sight is the behavior of Germany and Japan, who have little native energies and rush crazily into boom and bust economy on temporary and borrowed pipelines and tankers, throwing out what was stable and safe to become rich for a short period: monkey see, monkey do. Consider also Sweden, that once before boomed and busted in its age of Baltic Ships while cutting its virgin timber. Later it was completely stable on water power and agriculture, but now after a few years of growth became like the rest, another bunch of engines on another set of oil flows, a culture that may not be long for this world.

What is the general answer? Eject economic expansionism, stop growth, use available energies for cultural conversion to steady state, seek out the condition now that will come anyway, but by our service be our biosphere's handmaiden anew. ∎

Cosmic Economics *was an important contribution to the understanding of our energy use in relation to our economics—specifically the patterns of inflation. It was written when the authors were at the Oregon Governor's Office of Energy Research and Planning in 1974.*

COSMIC ECONOMICS

By Joel Schatz and Tom Bender

A CLEAR UNDERSTANDING of the fundamental relationships between energy, prices and inflation is essential before state, regional or national energy policies can be intelligently formulated.

Most of the fossil fuel energy that has powered our culture has come from concentrated and easily obtainable reserves. Now we must dig deeper, transport further, upgrade dilute energies (uranium, oil shale, etc.) to obtain our energy supply.

Although more total energy is produced each year, an increasing fraction of that energy is used up in obtaining the "net" energy available to the consumer. The consumer, in turn, must pay the cost of this increasing amount of "energy-getting energy" in addition to the energy cost of producing the goods and services he consumes. Everything which uses energy will cost more and more as net energy declines. This is the principal force driving world inflation.

At the same time that finite world energy reserves are being depleted, world demand and dependence upon them is accelerating. This greater competition for smaller and smaller reserves of energy is raising the monetary value of the remaining reserves, further increasing the price of energy. All the major new energy processes (oil shale,

nuclear, coal gasification, etc.) being developed to replace present fuels are even more costly than the fuels they are replacing, since they will require more energy and therefore more dollars to get the energy available to the consumer (i.e., they will generate even less net energy than traditional fuels).

Any energy policy which does not take net energy into consideration will bring about increasing economic instability. *The more successful the U.S. is in maintaining or increasing its total energy consumption,* UNDER CONDITIONS OF DECLINING NET ENERGY, *the more rapidly inflation, unemployment and general economic instability will increase. The disruptive effects of an inappropriate energy policy will be seen in terms of "economic crisis" rather than "energy crisis."*

Once this is recognized, the only prudent policy direction is to undertake an orderly transition away from exhaustible energy sources to inexhaustible energy sources (sun, wind, agriculture, tides, hydro, etc.) and to the level of consumption these sources can support.

While individual actions to conserve energy and materials are possible, and intrinsically worthwhile, it is unrealistic to expect such voluntary measures to occur on a large scale in our society. Unilateral conservation and slowdown on the part of individuals, states or regions should not be viewed, however, as self-sacrifice while others continue high-level consumption. Quite the contrary, those who attempt continued growth as net energies decline are merely creating the conditions for a sharper and more disruptive economic transition for themselves the longer they wait to adjust their consumption to what will inevitably be required of them. The most direct and least disruptive way to ease energy and economic transitions for society as a whole is through the enactment of two economic balancing mechanisms:

1. *A uniform tax levied on the potential energy content of all domestic exhaustible energy sources at the point of extraction*. The resulting tax revenue and increase in energy price will bring about the following:

• Allow us more time to make the transitions by slowing the depletion of our remaining exhaustible energy reserves.

• The use of our remaining exhaustible energy to finance and develop the structures and processes necessary to permit increased use of inexhaustible energy sources.

• The conversion of wasteful processes to lower and more efficient energy use (e.g. the development of mass transit systems).

• Equitable access to limited energy supplies, goods and services among all segments of society through use of energy tax revenues for income support for people on low or fixed incomes.

• The displacement of machines by human skills and labor, assuring full opportunity for employment.

• More efficient use of energy in our goods and services, lowering their cost of production and making them more competitive in foreign markets.

• Slower depletion of our domestic exhaustible energy reserves so they can be maintained as strategic stockpiles for emergency use, assuring our continuing economic, political and military independence.

• A lessening of the stresses placed on environmental systems by our industrial processes, reducing the energy and money required to prevent damage and restore vitality to our environment.

• Improvement in the performance and durability of buildings and manufactured goods.

2. *An extraction tax placed on the removal of all domestic raw materials from natural storage*. The resulting tax revenue and increasing price of extracted materials will facilitate the desired results of the uniform energy tax and will lead to the following additional changes:

• A marketplace incentive for replacing our open-ended organic and inorganic material flow with recycling processes, reducing the increasingly large amounts of energy needed to locate, concentrate and process raw materials.

• Encourage more efficient use of materials.

Implementation of a tax on energy and materials will eliminate piecemeal responses to our interrelated energy and economic problems and can relieve the need for many existing taxes.

Patterns of energy use inevitably adjust to the net energy levels available for goods and services. When net energy is increasing, it is necessary to expand total energy consumption, in order to maintain economic and social viability. Now, with net energy declining, the groundrules for energy use require us to lower our total energy consumption.

The primary mechanism of adjustment to declining net energy will be the slowing of energy consumption through accelerating prices. Although the proposed energy and materials extraction tax mea-

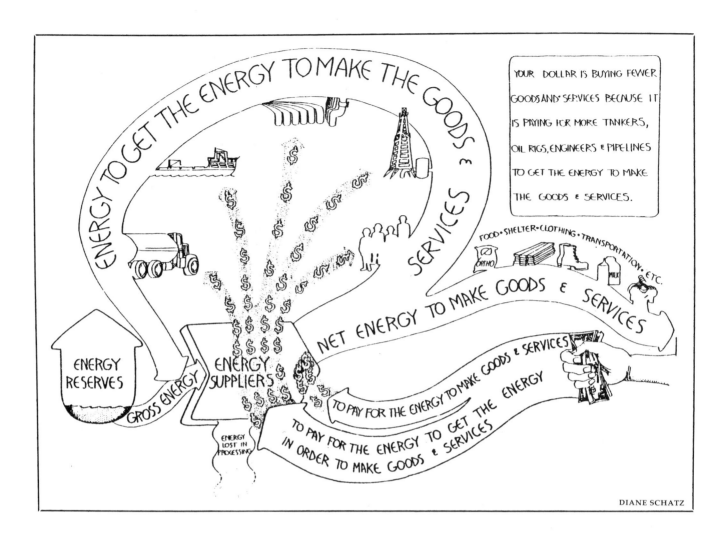

ENERGY TO GET THE ENERGY TO MAKE THE GOODS & SERVICES

YOUR DOLLAR IS BUYING FEWER GOODS AND SERVICES BECAUSE IT IS PAYING FOR MORE TANKERS, OIL RIGS, ENGINEERS & PIPELINES TO GET THE ENERGY TO MAKE THE GOODS & SERVICES.

FOOD·SHELTER·CLOTHING·TRANSPORTATION·ETC.

NET ENERGY TO MAKE GOODS & SERVICES

ENERGY RESERVES

GROSS ENERGY

ENERGY SUPPLIERS

ENERGY LOST IN PROCESSING

TO PAY FOR THE ENERGY TO MAKE GOODS & SERVICES

TO PAY FOR THE ENERGY TO GET THE ENERGY IN ORDER TO MAKE GOODS & SERVICES

DIANE SCHATZ

sures would ease this transition considerably, social and economic disruptions are inevitable. The severity of these disruptions will largely depend upon our ability to understand and accept the emerging requirements for stability.

It will also depend upon our capacity to see that the quality of our lives can actually improve as energy and materials prices increase. Lower energy use can lead to higher efficiency, greater opportunities for creative expression and individual freedom, and a stable climate for vigorous cultural evolution:

• Less unemployment, as the increasing energy costs of machines restore the value of human skill and labor.

• Better health, as sedentary and tension-producing work is replaced by more physically active work.

• Fresher and more nutritious food, as small-scale, localized and organic-based production replaces production processes dependent upon massive inputs of expensive fossil fuel energy and chemicals.

• Fewer accidents, as speeds are reduced in all of our activities.

• More worthwhile roles for older people as the need for their special skills and wisdom makes their contributions more valued.

• More integrated personal and family life, as work and family roles are less separated by space, time and organization.

• Less reliance on formal and abstract education separated from the rest of life, as high educational

28

costs stimulate processes such as apprenticeships where people are productive while learning, and as greater independence and self-respect renew ability to learn from oneself as well as from external standards.

• Higher quality surroundings, as the energy to disrupt large-scale environmental systems becomes unavailable.

• Stable prices and fewer inflationary pressures, as net energy levels stabilize.

• Less reliance upon credit buying, because of the increasing burden of its inflationary effects.

• More dignified death, as elaborate attempts to prolong and postpone inevitable processes become prohibitively expensive.

• Less commuting and less chauffeuring of family members as dispersed living patterns become too expensive.

• Less reliance on the use of all forms of drugs, as more fulfilling life-style options become available.

• Less guilt about future generations and other parts of the world as our capability of decreasing their options lessens.

• Less preoccupation with material goods and more with personal growth as our capacity for production of material goods approaches its limit and as we recognize the only direction infinite human desires can be channeled in a finite material world.

• More opportunity for craftsmanship and creativity, as independence and smaller scale of work organization gives more opportunity for self-directed and -controlled experimentation, resulting in more original and durable goods.

• More personal independence and freedom, as the higher costs of large-scale organization and economic and political centralization become prohibitive.

Those who are early to recognize the fundamental relationships between energy, prices and inflation will have an enormous advantage in moving toward a higher quality, lower energy way of life. ∎

THE RELATIVE DIFFICULTY of measuring qualitative things and the relative ease of measuring quantitative things has made it easier for us to value the quantities of life rather than its qualities. This has resulted in material growth becoming a fundamental value of our society, even after it ceases to improve the quality of our lives. Accepting the assumption that high energy and material use is necessary for a high quality of life has made it difficult for us to perceive the ease and benefit of reducing energy use.

Our actual needs for energy and material goods are quite small, and can be satisfied through inexhaustible energy sources and recycling of materials. Most of the things really important to our happiness and well-being have very little relationship to our use of energy—trust, a sense of belonging, love, identity, respect, accomplishment, harmony with our surroundings—and can only be gained through our personal growth, not simply the growth of our GNP. Our dreams are infinite, and attempts to realize them mainly through finite, limited material things can only lead to frustration and conflict.

The true value of our use of any exhaustible resource is not merely the enjoyment of its use, but its legacy. Responsible use of remaining non-renewable resources, whether domestic or imported, must insure that we survive their depletion and lead to new and better ways of life sustainable by a stable energy budget.

Tom Bender & Joel Schatz
From INDEPENDENCE?, 1974 (51)

30

VALUE AND VALUES

Much of the skepticism about the possibility of living with limited but carefully husbanded resources comes from those people who equate lack of growth with lack of progress. In other words, if a society is not producing more of everything all the time, it does not have a healthy economy. It is somehow underdeveloped.

Modern economists have been slow to perceive that not all growth is good and that growth in quantity is different from growth in quality. The nature and advisability of growth is different at different times in the lifetime of an individual or a civilization. As E. F. Schumacher said many times: When my children grow, I rest assured that all is well and I am feeding them the proper foods. If I grow, however, then I must start to pay attention to the amount and quality of the foods I am eating or I will grow too fat for my own good.

In economic terms, this means that there are appropriate levels of growth in any given economic situation. The stable or steady state is a natural counterbalance to an era of little or no regard for the nature or quality of its growth or for the health or long-term availability of the stores upon which it has been drawing.

In a collection of essays called Toward a Steady-State Economy *(18), economist Herman Daly develops some of the ideas which John Stuart Mill suggested during the 19th century in* The Principles of Political Economy *(41):*

> *By "steady state" is meant a constant stock of physical wealth (capital) and a constant stock of people (population). Naturally these stocks do not remain constant by themselves. People die, and wealth is physically consumed—that is, worn out, depreciated. Therefore, the stocks must be maintained by a rate of inflow (birth, production) equal to the rate of outflow (death, consumption).*

A steady state can happen when a society begins to make the transition from "high production efficiency" to "high maintenance efficiency." Daly borrows these terms, which Eugene P. Odum uses to describe eco-system succession, and applies them to the human economy. Young, wealthy economies, like young ecosystems, maximize production, growth and quantity. Mature, steady-state economies stress protection, stability and quality.

> *Most economists are still hung up on the assumption of infinite wants, or the postulate of nonsatiety as the mathematical economists call it. Any single want can be satisfied, but all wants in the aggregate cannot be. Wants are infinite in number if not in intensity, and the satisfaction of some wants stimulates other wants. If wants are infinite, growth is always justified—or so it would seem.*
>
> *Even while accepting the foregoing hypothesis, one could still object to growthmania on the grounds that, given the completely inadequate definition of GNP, "growth" simply means the satisfaction of ever more trivial wants, while simultaneously creating ever more powerful externalities which destroy ever more important environmental amenities. To defend ourselves against these externalities we produce even more, and instead of subtracting the purely defensive expenditures we add them! For example, the medical bills paid for treatment of cigarette-induced cancer and pollution-induced emphysema are added to GNP when in a welfare sense they should clearly be subtracted. This should be labeled "swelling," not "growth." Also the satisfaction of wants created by brainwashing and "hog-washing" the public over the mass media represent mostly swelling. A policy of maximizing GNP is practically equivalent to a policy of maximizing depletion and pollution.*
>
> *The steady state would make fewer demands on our environmental resources, but much greater demands on our moral resources. In the past a good case could be made that leaning too heavily on scarce moral resources, rather than relying on abundant self-interest, was the road to serfdom. But in*

an age of rockets, hydrogen bombs, cybernetics, and genetic control, there is simply no substitute for moral resources, and no alternative to relying on them, whether they prove sufficient or not.

Since traditional Western economics does not measure anything that is not clearly marked with a price tag—or contribute to the Gross National Product—part of the transition towards a stable state involves recognizing factors which are essential to a healthy economy but are difficult to quantify. This means giving economic value to the natural processes and organisms that have been quietly working all along to purify and sustain the environment. Salt marshes, for example, have been systematically destroyed although we now know they are the source of micro-organisms which nourish crab, fish and other sea creatures that do have economic value for the fishing industry. Estuaries are counted only when they have been filled in for housing or industrial development. By the same token, trees have economic value only when they are cut and sold for timber, but not when they stand, providing shade and oxygen in a hot, polluted city. In The Clothesline Paradox, reprinted here, Steve Baer points out what the sun has been doing for us that we have neglected to take into account.
Mutual aid and self-support systems which do not involve a cash transaction are also ignored. Nearly 50 percent of U.S. families grow vegetables in a backyard or community garden. None of this shows up in indices of agricultural productivity or in the GNP, though the products of the garden could substantially support and sustain a family, freeing it from the supermarket economy. In his book, The Household Economy, Scott Burns looks at the real productive value of household economic activity. Our mistake, he says, has been to consider the household as a mere consumer of goods. The result has been to ignore a major portion of a lively and strong center of economic activity.

An understanding of what (or whether) a society measures or counts is one clue to understanding where its values lie. As we shift from a no-holds-barred growth economy, realizing that the earth can no longer sustain such growth, we must take a closer look at the values inherent in any course of action we may take. Nothing is value-free—everything we do moves us towards certain ends. We need to be conscious of these ends and choose the means which serve ends we value.

An even deeper perspective on our basic cultural values can be gained by turning 180 degrees from Western culture to examine some traditional Eastern values and how they are reflected in their economies. As C. R. Ashbee inferred in his essay at the beginning of this book, the Eastern ethic is based upon a sustainable, high quality state where craftsmanship, caring and the careful use of resources are rewarded. In Buddhist Economics, E. F. Schumacher contrasts the trade-offs and ramifications of this way of thinking to those of the fast-moving, technological world of the West. The values are not necessarily ones we need to adopt, but they are ones worth exploring as more appropriate to the times we are moving into.

What is wealth? What is affluence? When is too much of a good thing? What is enough? In Conscious Culture of Poverty, Schumacher talks about choosing the kind of wealth that comes from frugal, careful living—properly discriminating between things that have eternal value and those which are rightly used up and gone. He provides here another mirror into the upside-down, inside-out and backwards ways we have been operating.

Many of the values and patterns Schumacher emphasizes are the traditional, deeply felt values of our own culture—ones we have been paying only lip service to while nations, corporations and individuals scramble for larger and larger pieces of the pie. They are, in fact, the common sense ethics which Americans pride themselves in. In this second excerpt from Sharing Smaller Pies, Tom Bender outlines briefly some of these values—both new and old, American and Buddhist— which we need to be paying attention to.□

Steve Baer is an inventor and author who is an old hand at poking holes in heretofore accepted thinking. This piece rang a good many bells when it first appeared in the Summer, 1975, issue of the New Mexico Solar Energy Society Newsletter *(43). Reprinted with permission from the author.*

THE CLOTHESLINE PARADOX

By Steve Baer

A FEW YEARS AGO Peter Van Dresser mentioned the Clothesline Paradox.

Solar energy advocates are continuously humiliated by being shown "energy pies." Slices are assigned to coal, gas, oil, hydroelectric and even nuclear, but solar energy is evidently too small to appear. I have a typical energy pie from the Ford Foundation whose source is the U.S. Bureau of Mines. The large pie is split into five pieces. Petroleum 46%, coal 18%, natural gas 31%, hydropower 4%, and nuclear 1%. (An asterisk notes that wood has been omitted—why?) We are frequently reminded that the energy we advocate—solar energy —must, after the proper technical efforts, appear alongside coal, oil, natural gas and nuclear before it will make an "impact." ERDA in its different energy consumption predictions assigns only a thin wedge of the pie to solar energy and then only as a faint hope 15 to 25 years from now. The demoralized reader is then ripe to be persuaded of the necessity of nuclear power plants or offshore drilling. The accounting system shows that he has done absolutely nothing with solar energy. He lacks even a trace of a useful habit or activity that he could build on. As Peter and I discussed—if you examine these figures you find the cards are stacked against solar energy.

If you take down your clothesline and buy an electric clothes dryer, the electric consumption of the nation rises slightly. If you go in the other direction and remove the electric clothes dryer and install a clothesline, the consumption of electricity drops slightly, but there is no credit given anywhere on the charts and graphs to solar energy, which is now drying the clothes.

The poor old sun is badly mistreated by such graphs. In the first place, the obvious should be pointed out: that coal, oil and natural gas are all solar energy products stored ages ago by photosynthesis, and hydroelectric power is solar energy no older than the weather patterns which dropped the precipitation flowing through the turbines.

The graphs which demonstrate a huge dependence on fossil fuels are fine in one respect. They are alarming. But they are very bad in another respect. They are misleading. Misleading to such an extent that they blind people to obvious answers and prime them to a frenzy of effort in poor directions. Attention given to such graphs and charts trains people to attempt to deliver what is shown in these accounting systems rather than what is needed.

If you drive a motorcycle, the gasoline you consume appears in the nation's energy budget. If you get a horse to ride and graze the horse on range nearby, the horse's energy which you use does not appear in anyone's energy accounting.

If you install interior greenhouse lights, the electricity you use is faithfully recorded. If you grow the plants outside, no attempt is made at an accounting.

If you drive your car to the corner to buy a newspaper, the gasoline consumption appears. If you walk—using food energy—the event has disappeared from sight, for the budget of solar energy consumed by people in food is seldom mentioned.

The Ford Foundation's energy study shows the U.S.'s energy consumption in 1968 at about 62 quadrillion Btu, or 310,000,000 Btu/person/year,

or 310/365=850,000 Btu/day. If the average daily caloric intake is 2500 Kcal., this is approximately 10,000 Btu/day/person—about 1.2% of the total consumption listed by the Bureau of Mines. But this 1.2% doesn't appear anywhere on the graphs. Nuclear energy with 1% does appear. The food is obviously solar energy. Why is it not included?

What about the question of the energy used in growing the food? Can't we treat this in the same way as the coal burned to generate electricity? If we use the figure of .5% efficiency (Ayres and Scarlott), this means we have consumed approximately 2,000,000 Btu/person/day of sunlight in producing the 10,000 Btu/person consumed. Solar energy then immediately fills over ⅔ of the new energy pie. If we aren't allowed to show the actual sunlight required for our 10,000 Btu/person, then what about power plants? Why is it that when they burn 4 Btu of fuel for every Btu delivered as electricity all the consumption appears in the energy accounts rather than the 1 Btu?

Why wouldn't it be fair to expand the slice—4% (1973, Bureau of Mines) given to hydroelectric power by a similar factor of efficiency—for the solar energy consumed in raising the water to its working head? After all, in most cases, the rain or snow fell through long unexploited distances before it went to work in a power plant.

Then there is the question of heating houses. Every time the sun shines on the surface of a house, and especially when it shines through a window, there is "solar heating" to some extent. How do we measure this? How do we account for this in our discussions of energy use? According to the NSF/NASA Energy Panel of 1972, the percentage of thermal energy for buildings supplied by the sun was too small to be measurable. But is that accurate? Shouldn't we recalculate the energy consumption of every building, assuming it were kept in the shade all day, and then attribute the difference between this amount and its actual consumption to solar energy? In most cases this would result in an enormous difference. Almost every building is solar heated to some extent. I would guess the average shaded fuel consumption to be at least 15% higher, and then of course our next concern in heating the building is what keeps the earth as warm as it is? What supplies the United States with the necessary energy to maintain an average temperature of 60 degrees Fahrenheit as it spins in empty space at absolute zero? This is a heating contract no oil company would be quick to try and fill.

Clearly it would be a very difficult thing to account for every calorie or Btu that passed through us or by us every day in the various forms. It doesn't seem to be a particularly urgent job, but it is very important to examine what the limits of an accounting system are—to know what the numbers and quantities displayed really mean.

If you go to a drive-in movie to watch the flickering lights on a screen, the energy consumption of the automobile and the drive-in is dutifully recorded and appears in the statistics. If you walk out on a hillside, lie on your back and look at the stars, no attempt is made to measure the power output of the distant stars.

I don't advocate an enormous effort to measure all these things. It would just be more helpful if the graphs stated more clearly what they are about.

The design of houses can be stilted by such graphs. Now that the experts have started this infantile accounting system, which evidently finds us completely independent of the sun, solar energy will be admitted only so long as it has been properly collected, stored and transferred. Legislation aimed at encouraging the use of solar energy equipment by subsidizing the price of certain hardware must end by being pathetic and blundering. It would take an enormous crew of experts to determine the efficiency of different orientations of windows, different arrangements of shade trees, etc., etc. To ignore these efforts and only to reward the purchase of "off the shelf hardware" is to further the disease of narrow-minded quantification.

It should be pointed out to the people promoting the use of solar energy in the place of fossil fuels that the accounting systems used by the experts are rigged against them. As I understand it, we are being prepared to accept that there are legitimate and illegitimate ways of using the sun. If you purchase certain kinds of hardware to exploit solar energy, it will be accounted for and a credit will be given to the sun. If you depend on more customary old-fashioned uses of solar energy—growing food, drying clothes, sun bathing, warming a house with south windows, the sun credit is totally ignored.

Our present accounting system, with its promise of a credit to the sun after the right hardware has been installed, can only discourage good house design. If the natural solar contribution to house heating from windows is ignored, then the designer

knows that expanding this share done by the sun will also be ignored. No tax incentives—no credit given to the sun in ERDA's graphs.

I think we would be much better informed if alongside every graph showing our use of oil, coal and uranium there were also an indication of the total energy received from the sun. Since we can't do without it, let's not omit it from our accounts. In the case of the United States, a conservative estimate of the solar energy received in one year might be:

$$(3,000,000) \text{ sq. mi. } (5280^2)\text{ft.}^2)\text{mi.}^2 \text{ x}$$
$$350 \text{ x } 10^3 \text{ Btu received/ft}^2)\text{year} = 293 \text{ x } 10^{17} \text{ Btu/yr.}$$

Twenty-nine thousand, three hundred quadrillion Btu, as opposed to the 62 quadrillion shown as used during 1968 by the U.S. Bureau of Mines.

When small children first start paying close attention to money and to their allowances, they briefly commit their whole minds to their few coins and what chores they did to earn them—without even considering the budget of the family's household. We can't allow our entire civilization to be similarly ignorant for long. We must ask who's keeping score and why they have such peculiar methods. ∎

The full text of Household Economy *(13)—originally published as* Home Inc. *in 1975—explains in greater detail the ideas presented here by Scott Burns. His current writings appear in an economics column in the* Boston Herald. *Reprinted with permission from Doubleday.*

THE HOUSEHOLD ECONOMY

By Scott Burns

From THE HOUSEHOLD ECONOMY

THE IDEA that the household is a productive economic institution is probably a novel one to most readers. We are educated to consider the household as a *consuming* institution, dedicated to absorbing that which is produced by the marketplace. At best, we are likely to think of the family as a means for transferring traditions, values and habits from one generation to the next; at worst, we are inclined to think of it as a vestigial institution whose continued existence is painful and embarrassing.

But the household *is* a productive economic institution. It produces goods and services with a tangible economic value. Like the market economy, the household economy employs labor and capital and strives to increase the benefits that accrue to its owners and managers. The main reason that we have failed to perceive the household as a productive institution is that it was, until recently, relatively less important than the marketplace. This, as we shall see, is no longer the case.

Another reason we have failed to perceive the household accurately is that it is peculiarly handicapped. It has no means of exchange; thus it employs labor but pays no cash wages, and it invests in capital goods but issues no cash dividends. It uses no money to measure its production of goods and services. Its economic product is accounted for by transfers *in kind*, within the family proper. It has, quite simply, been defined out of existence by the conventions of economic accounting and is assured of remaining officially invisible for some time to come. In the market economy, only cold cash counts, and the market economy, for the moment, directs and defines conventional perceptions.

What *is* the household economy? It is the sum of all the goods and services produced within all the households in the United States. This includes,

among other things, the value of shelter, home-cooked meals, all the weekend-built patios and barbecues in suburban America, painting and wallpapering, home sewing, laundry, child care, home repairs, volunteer services to community and to friends, the produce of the home garden, and the transportation services of the private automobile.

There were, according to the 1970 Census, more than 51 million family households in the United States. More than one million new families are created every year. Each employs labor. Although the vast bulk of the labor employed is provided by housewives, significant contributions are also made by husbands and children. Some research surveys have indicated that the total amount of time devoted to household labor is increasing rather than decreasing. Similarly, each household invests in, and accumulates, a stock of capital goods that are distinct and separate from the market economy. These goods include the house, car, household appliances and TV set, as well as the lawn mower, storm windows, power drill and portable saw. The return on these investments is measured by the services rendered rather than in dollars; thus, it fails to appear in conventional statistics of national product or in orthodox economic teaching. While all these goods are usually considered the useless paraphernalia of a passive consumer society or the concrete burden of a bourgeois life, they are, in fact, the productive capital of a vital and very private economy.

While it is difficult to imagine surviving, let alone sustaining, most of what we now consider to be the good life, without the organization and output of the market economy, the household economy is, in fact, quite capable of producing many of the goods and services now produced by the market. The household economy of the future, like the household economy of the past, will be a central and creative force, rather than an inconvenient institution organized to perform the tasks not yet taken on by the marketplace.

• • •

How large would this invisible economy be if it could be measured in dollars? *Very* large. According to one study, the total value of all the goods and services produced by the household economy in 1965 was about $300 *billion*. This was about equal to the gross national product of the Soviet Union at that time. If all the work done within the household by men and women were monetized, the total would be equal to the entire amount paid out in wages and salaries by every corporation in the United States. Similarly, the assets commanded by households, *worth more than a trillion dollars*, produce an annual return in goods and services almost equal to the net profits of every corporation in the United States. Very, very little of this appears in conventional accountings for the gross national product.

This neglect is rather like assuming, say, that the entire European Free Trade Association (Austria, Denmark, Norway, Portugal, Sweden, Switzerland and the United Kingdom) has no economic product. It is also like assuming that somewhere between a quarter and a third of our total economic product does not exist—a large lump to hide under any rug.

The invisible household economy might also be called the matriarchal economy, because it is dominated by women. They perform most of the labor, make most of the household decisions, and are employed as managers for the labor and assets of the household. More than a few observers have noted that the household economy is invisible precisely *because* it is controlled by women and that present accounting conventions have the effect of demeaning the work and value of women. The economist or sociologist who allows his perception of the world to be circumscribed by the measures of the market economy—as most do—allows his ideas, observations and theories to be dictated by the limits of the marketplace, thereby excluding a legitimate third of our economic activity and most of the activity of women. His habit of thought is repeated in the larger world.

In the composition of the national income we learn that the unpaid work of mothers and housewives is excluded. This is not, as we shall soon see, a trifling matter. Not only does it work to discriminate against recognizing the economic contribution of housewives, it also distorts our perception of economic growth.

Although women are the chief victims of this exclusion, they are not alone. Virtually all work that is not rewarded with wages is excluded from conventional accounts of national income. While the household economy is, by far, the largest omission, the volunteer economy and the cooperative economy are also excluded. The common denominator of these forms of economic activity, beyond their failure to use money, is that they are organized around the idea of giving, of mutual need, and of cooperation rather than competition. They assume that productive activity is a social as well as an economic function and that the competitive drive for individual gain is not necessarily the best drive to harness in order to accomplish a given task. Perversely, our system of economic accounts excludes all motives but the competitive desire for money.

In the past, when the productive activity of the household was being drawn into the market economy by the industrial revolution, the rate and amount of true economic growth was overstated. Economic activity was not created, it was merely *transferred* from the household (where it was not counted) to the market economy (where it was). The labor of the woman who "put up" fruits and vegetables for the winter by home canning was not included in national income. When she took a job in a canning factory and purchased canned fruit and vegetables in a store, both her work and that of the clerk who sold her the goods became part of the national income. Its growth was applauded irrespective of its illusory nature.

Now similar inaccuracies are being created in reverse. As this is written, it is virtually impossible to buy home canning equipment, because the demand has run so far ahead of the supply. The Ball Corporation, the nation's oldest and largest supplier of home canning equipment, reports a surge in sales and a consumer interest dominated by the young. While their record sales will be reflected in the national income accounts, the work of canning done in the home implied by those sales will be excluded. Nor will the labor that went into home gardens or home sewing be considered, or the consumer's labor at self-service gas stations and food cooperatives. Where once the conventions of national income accounting worked to overstate the rate of true economic growth, they may now understate it, as work is created for the household economy but not recognized. ■

Buddhist Economics *was the first of E. F. Schumacher's essays to catch people's eyes and is still the very best statement about values, work and livelihood. It was first published in 1966, in* Asia: A Handbook, *edited by Guy Wint (62). The first U.S. printing was in* Manas *(38). Since then it has appeared in many collections, including* Small Is Beautiful *(52). It cannot be read too many times. Reprinted with permission from* Resurgence *(50).*

BUDDHIST ECONOMICS

By E. F. Schumacher

RIGHT LIVELIHOOD is one of the requirements in the Buddha's Noble Eightfold Path. It is clear, therefore, that there must be such a thing as Buddhist Economics.

Buddhist countries, at the same time, have often stated that they wish to remain faithful to their heritage. So Burma: "The New Burma sees no conflict between religious values and economic progress. Spiritual health and material well-being are not enemies: they are natural allies." Or "We Burmans have a sacred duty to conform both our dreams and our acts to our faith. This we shall ever do."

All the same, such countries invariably assume that they can model their economic development plans in accordance with modern economics, and they call upon modern economists from so-called advanced countries to advise them, to formulate the policies to be pursued, and to construct the grand design for development, the Five-Year Plan or whatever it may be called. No one seems to think that a Buddhist way of life would call for Buddhist economics just as the modern materialist way of life has brought forth modern economics.

Economists themselves, like most specialists, normally suffer from a kind of metaphysical blindness, assuming that theirs is a science of absolute and invariable truths, without any presuppositions. Some go as far as to claim that economic laws are as free from "metaphysics" or "values" as the law of gravitation. We need not, however, get involved in arguments of methodology. Instead, let us take some fundamentals and see what they look like when viewed by a modern economist and a Buddhist economist.

There is a universal agreement that the fundamental source of wealth is human labor. Now, the modern economist has been brought up to consider "labor" or work as little more than a necessary evil. From the point of view of the employer, it is in any case simply an item of cost, to be reduced to a minimum if it cannot be eliminated altogether, say, by automation. From the point of view of the workman, it is a "disutility"; to work is to make a sacrifice of one's leisure and comfort, and wages are a kind of compensation for the sacrifice. Hence, the ideal from the point of view of the employer is to have output without employees, and the ideal from the point of view of the employee is to have income without employment.

The consequences of these attitudes, both in theory and in practice, are, of course, extremely far-

reaching. If the ideal with regard to work is to get rid of it, every method that "reduces the work load" is a good thing. The most potent method, short of automation, is the so-called "division of labor" and the classic example is the pin factory eulogized in Adam Smith's *Wealth of Nations*. Here it is not a matter of ordinary specialization, which mankind has practiced from time immemorial, but of dividing up every complete process of production into minute parts, so that the final product can be produced at great speed without anyone having had to contribute more than a totally insignificant and, in most cases, unskilled movement of his limbs.

The Buddhist point of view takes the function of work to be at least threefold: to give a man a chance to utilize and develop his faculties; to enable him to overcome his ego-centeredness by joining with other people in a common task, and to bring forth the goods and services needed for a becoming existence. Again, the consequences that flow from this view are endless. To organize work in such a manner that it becomes meaningless, boring, stultifying, or nerve-racking for the worker would be little short of criminal; it would indicate a greater concern with goods than with people, an evil lack

of compassion and a soul-destroying degree of attachment to the most primitive side of this worldly existence. Equally, to strive for leisure as an alternative to work would be considered a complete misunderstanding of one of the basic truths of human existence, namely that work and leisure are complementary parts of the same living process and cannot be separated without destroying the joy of work and the bliss of leisure.

From the Buddhist point of view, there are therefore two types of mechanization which must be clearly distinguished: one that enhances a man's skill and power and one that turns the work of man over to a mechanical slave, leaving man in a position of having to serve the slave. How to tell one from the other? "The craftsman himself," says Ananda Coomaraswamy, a man equally competent to talk about the Modern West as the Ancient East, "the craftsman himself can always, if allowed to, draw the delicate distinction between the machine and the tool. The carpet loom is a tool, a contrivance for holding warp threads at a stretch for the pile to be woven round them by the craftsman's fingers; but the power loom is a machine, and its significance as a destroyer of culture lies in the fact that it does the essentially human part of the work." It is clear, therefore, that Buddhist economics must be very different from the economics of modern materialism, since the Buddhist sees the essence of civilization not in a multiplication of wants but in the purification of human character. Character, at the same time, is formed primarily by a man's work. And work, properly conducted in conditions of human dignity and freedom, blesses those who do it and equally their products. The Indian philosopher and economist J. C. Kumarappa sums the matter up as follows:

> If the nature of the work is properly appreciated and applied, it will stand in the same relation to the higher faculties as food is to the physical body. It nourishes and enlivens the higher man and urges him to produce the best *he is capable of. It directs his freewill along* the proper course and disciplines the animal in him into progressive channels. It furnishes an excellent background for man to display his scale of values and develop his personality.

If a man has no chance of obtaining work, he is in a desperate position, not simply because he lacks an income but because he lacks this nourishing and enlivening factor of disciplined work which nothing can replace. A modern economist may engage in highly sophisticated calculations on whether full employment "pays" or whether it might be more "economic" to run an economy at less than full employment so as to ensure a greater mobility of labor, a better stability of wages, and so forth. His fundamental criterion of success is simply the total quantity of goods produced during a given period of time. "If the marginal urgency of goods is low," says Professor Galbraith in *The Affluent Society*, "then so is the urgency of employing the last man or the last million men in the labor force." And again:

> If . . . we can afford some unemployment in the interest of stability—a proposition, incidentally, of impeccably conservative antecedents—then we can afford to give those who are unemployed the goods that enable them to sustain their accustomed standard of living.

From a Buddhist point of view, this is standing the truth on its head by considering goods as more important than people and consumption as more important than creative activity. It means shifting the emphasis from the worker to the product of work, that is, from the human to the sub-human, a surrender to the forces of evil. The very start of Buddhist economic planning would be a planning for full employment, and the primary purpose of this would in fact be employment for everyone who needs an "outside" job: it would not be the maximization of employment nor the maximization of production. Women, on the whole, do not need an "outside" job, and the large-scale employment of women in offices or factories would be considered a sign of serious economic failure. In particular, to let mothers of young children work in factories while the children run wild would be as uneconomic in the eyes of a Buddhist economist as the employment of a skilled worker as a soldier in the eyes of a modern economist.*

While the materialist is mainly interested in goods, the Buddhist is mainly interested in liberation. But Buddhism is "The Middle Way" and therefore in no way antagonistic to physical well-being. It is not wealth that stands in the way of liberation but the attachment to wealth—not the enjoyment of pleasurable things but the craving for them. The keynote of Buddhist economics, therefore, is simplicity and non-violence. From an economist's point of view, the marvel of the Buddhist way of life is the utter rationality of its pattern—amazingly small means leading to extraordinarily satisfactory results.

For the modern economist this is very difficult to understand. He is used to measuring the "standard of living" by the amount of annual consumption, assuming all the time that a man who consumes more is "better off" than a man who consumes less. A Buddhist economist would consider this approach excessively irrational—since consumption is merely a means to human well-being, the aim should be to obtain the maximum of well-being with the minimum of consumption. Thus, if the purpose of clothing is a certain amount of temperature comfort and

---

*This line started a controversy—one which Fritz Schumacher himself perhaps never took seriously enough due to his "old world" values and culture. In many traditional and colonial economies, it is a particularly vicious form of exploitation if women work while men stay home. Women are offered the jobs because they can be paid so much less. Most of them have no desire to be working under excruciating factory or field conditions. At the same time, the men are shamed by the relegation from the role of breadwinner to idleness. (In such situations, few men do the housework or childtending—that is more often left to older children or grandmothers.) In our society Schumacher's point no longer holds true. Indeed, part of the re-examination of values so important in changing times is the realization that both sexes are more fulfilled when work in the home and in the "outside" world is shared and balanced. (Ed. note)

an attractive appearance, the task is to attain this purpose with the smallest possible effort; that is, with the smallest annual destruction of cloth and with the help of designs that involve the smallest input of toil. The less toil there is, the more time and strength is left for artistic creativity. It would be highly uneconomic, for instance, to go in for complicated tailoring, like the modern West, when a much more beautiful effect can be achieved by the skillful draping of uncut material. It would be the height of folly to make anything ugly, shabby or mean. What has just been said about clothing applies equally to all other human requirements. The ownership and the consumption of goods is a means to an end, and Buddhist economics is the systematic study of how to attain given ends with the minimum means.

Modern economics, on the other hand, considers consumption to be the sole purpose of all economic activity, taking the factors of production—land, labor and capital—as the means. The former, in short, tries to maximize human satisfactions by the optimal pattern of productive effort. It is easy to see that the effort needed to sustain a way of life which seeks to attain the optimal pattern of consumption is likely to be much smaller than the effort needed to sustain a drive for maximum consumption. We need not be surprised, therefore, that the pressure and strain of living is very much less in, say, Burma than it is in the United States, in spite of the fact that the amount of labor-saving machinery used in the former country is only a minute fraction of the amount used in the latter.

Simplicity and non-violence are obviously closely related. The optimal pattern of consumption, producing a high degree of human satisfaction by means of a relatively low rate of consumption, allows people to live without great pressure and strain and to fulfill the primary injunction of Buddhist teaching: "Cease to do evil; try to do good." As physical resources are everywhere limited, people satisfying their needs by means of a modest use of resources are obviously less likely to be at each other's throats than people depending upon a high rate of use. Equally, people who live in highly self-sufficient local communities are less likely to get involved in large-scale violence than people whose existence depends on world-wide systems of trade.

From the point of view of Buddhist economics, therefore, production from local resources for local needs is the most rational way of economic life,

while dependence on imports from afar and the consequent need to produce for export to unknown and distant peoples is highly uneconomic and justifiable only in exceptional cases and on a small scale. Just as the modern economist would admit that a high rate of consumption of transport services between a man's home and his place of work signifies a misfortune and not a high standard of life, so the Buddhist economist would hold that to satisfy human wants from far-away sources rather than from sources nearby signifies failure rather than success. The former might take statistics showing an increase in the number of ton/miles per head of the population carried by a country's transport system as proof of economic progress, while to the latter— the Buddhist economist—the same statistics would indicate a highly undesirable deterioration in the pattern of consumption.

Another striking difference between modern economics and Buddhist economics arises over the use of natural resources. Bertrand de Juvenal, the eminent French political philosopher, has characterized "Western man" in words which may be taken as a fair description of the modern economist:

> He tends to count nothing as an expenditure, other than human effort; he does not seem to mind how much mineral matter he wastes and, far worse, how much living matter he destroys. He does not seem to realize at all that human life is a dependent part of an ecosystem of many different forms of life. As the world is ruled from towns where men are cut off from any form of life other than human, the feeling of belonging to an ecosystem is not revived. This results in a harsh and improvident treatment of things upon which we ultimately depend, such as water and trees."

The teaching of the Buddha, on the other hand, enjoins a reverent and non-violent attitude not only to all sentient beings, but also, with great emphasis, to trees. Every follower of the Buddha ought to plant a tree every few years and look after it until it is safely established, and the Buddhist economist can demonstrate without difficulty that the universal observance of this rule would result in a high rate of genuine economic development independent of any foreign aid. Much of the economic decay of Southeast Asia (as of many other parts of the world) is undoubtedly due to a heedless and shameful neglect of trees.

Modern economics does not distinguish between renewable and non-renewable materials, as its very method is to equalize and quantify everything by means of a money price. Thus, taking various alternative fuels, like coal, oil, wood or water power: the only difference between them recognized by modern economics is relative cost per equivalent unit. The cheapest is automatically the one to be preferred, as to do otherwise would be irrational and "uneconomic."

From a Buddhist point of view, of course, this will not do; the essential difference between non-renewable fuels like coal and oil on the one hand and renewable fuels like wood and water power on the other cannot be simply overlooked. Non-renewable goods must be used only if they are indispensable, and then only with the greatest care and the most meticulous concern for conservation. To use them heedlessly or extravagantly is an act of violence, and while complete non-violence may not be attainable on this earth, there is nonetheless an ineluctable duty on man to aim at the ideal of non-violence in all he does.

Just as a modern European economist would not consider it a great economic achievement if all European art treasures were sold to America at attractive prices, so the Buddhist economist would insist that a population basing its economic life on non-renewable fuels is living parasitically, on capital instead of income. Such a way of life could have no permanence and would therefore be justified only as a purely temporary expedient. As the world's resources of non-renewable fuels—coal, oil and natural gas—are exceedingly unevenly distributed over the globe and undoubtedly limited in quantity, it is clear that their exploitation at an ever-increasing rate is an act of violence against nature which must inevitably lead to violence between men.

This fact alone might give food for thought even to those people in Buddhist countries who care nothing for the religious and spiritual values of their heritage and ardently desire to embrace the materialism of modern economics at the fastest possible speed. Before they dismiss Buddhist economics as nothing better than a nostalgic dream, they might wish to consider whether the path of economic development outlined by modern economics is likely to lead them to places where they really want to be. Towards the end of his courageous book, *The Challenge of Man's Future*, Professor Harrison Brown of the California Institute of Technology gives the following appraisal:

> Thus we see that, just as industrial society
> is fundamentally unstable and subject to reversion to
> agrarian existence, so within it the conditions which

offer individual freedom are unstable in their ability to avoid the conditions which impose rigid organization and totalitarian control. Indeed, when we examine all of the foreseeable difficulties which threaten the survival of industrial civilization, it is difficult to see how the achievement of stability and the maintenance of individual liberty can be made compatible.

Even if this were dismissed as a long-term view—and in the long term, as Keynes said, we are all dead—there is the immediate question of whether "modernization" as currently practiced without regard to religious and spiritual values, is actually producing agreeable results. As far as the masses are concerned, the results appear to be disastrous—a collapse of the rural economy, a rising tide of unemployment in town and country, and the growth of a city proletariat without nourishment for either body or soul.

It is in the light of both immediate experience and long-term prospects that the study of Buddhist economics could be recommended even to those who believe that economic growth is more important than any spiritual or religious values. For it is not a question of choosing between "modern growth" and "traditional stagnation." It is a question of finding the right path of development, the Middle Way between materialist heedlessness and traditionalist immobility, in short, of finding "Right Livelihood."

That this can be done is not in doubt. But it requires much more than blind imitation of the materialist way of life of the so-called advanced countries. It requires above all the conscious and systematic development of a Middle Way in technology, of an "intermediate technology," as I have called it, a technology more productive and powerful than the decayed technology of the ancient East, but at the same time non-violent and immensely cheaper and simpler than the labor-saving technology of the modern West. ∎

When Conscious Culture of Poverty *first appeared in* Resurgence *(50) in March, 1975, it struck a strong chord in those of us who were trying to get a handle on the roots of the problems we were dealing with in U.S. culture. Publisher of many of Schumacher's essays,* Resurgence *is a good place to find new ideas. A collection of their best is now available in* Time Running Out? *(44), edited by Michael North. Reprinted with permission from* Resurgence.

CONSCIOUS CULTURE OF POVERTY

By E. F. Schumacher

Schon in der Kindheit hort' ich es mit Beben:
*Nur wer im Wohlstand lebt, lebt angenehm.**
—BERTOLT BRECHT

ONLY THE RICH can have a good life. This is the daunting message that has been drummed into the ears of all humankind during the last half-century or so. It is the implicit doctrine of "development;" the growth of income serves as the very criterion of progress. Everyone, it is held, has not only the right but the duty to become rich, and this applies to societies even more stringently than to individuals. The most succinct and most relevant indicator of a country's status in the world is thought to be *average income per head*, while the prime object of admiration is not the level already attained, but the current *rate of growth*.

It follows logically—or so it seems—that the greatest obstacle to progress is a growth of population: it frustrates, diminishes, offsets what the growth of Gross National Product (GNP) would otherwise achieve. What is the point of, let us say, doubling GNP over a period if population is also allowed to double during the same time? It would mean running fast merely to stand still: *average income per head* would remain stationary, and there would be no advance at all towards the cherished goal of universal affluence.

In the light of this received doctrine, the well-nigh unanimous prediction of the demographers—that

*In unpoetical English: "Even as a child I felt terror-struck when I heard it said that to live an agreeable life you have got to be rich."

world population, barring unforeseen catastrophes, will double during the next thirty years—is taken as an intolerable threat. What other prospect is this than one of limitless frustration?

Some mathematical enthusiasts are still content to project the economic "growth curves" of the last thirty years for another thirty or even fifty years, to "prove" that all humankind can become immensely rich within a generation or two. Our only danger, they suggest, is to succumb, at this glorious hour in the history of progress, to a "failure of nerve." They presuppose the existence of limitless resources in a finite world; an equally limitless capacity of living nature to cope with pollution; and the omnipotence of science and social engineering.

The sooner we stop living in the cloud-cuckoo-land of such fanciful projections and presuppositions, the better it will be, and this applies to the people of the rich countries just as much as to those of the poor. *It would apply even if all population growth stopped entirely forthwith.*

The modern assumption that "only the rich can have a good life" springs from a crudely materialistic philosophy which contradicts the universal tradition of humankind. The material *needs* of human beings are limited and in fact quite modest, even though our material *wants* above our needs can give us the "good life."

POVERTY IS NOT MISERY

To make my meaning clear, let me state right away that there are degrees of poverty which may be totally inimical to any kind of culture in the ordinarily accepted sense. They are essentially different from "poverty" and deserve a separate name; the term that offers itself is *misery*. We may say that poverty prevails when people have enough to keep body and soul together but little to spare, whereas in misery they cannot keep body and soul together, and even the soul suffers deprivation. Some thirteen years ago, when I began seriously to grope for answers to these perplexing questions, I wrote this in "Roots of Economic Growth"**

"All peoples—with exceptions that merely prove the rules—have always known how to help themselves; they have always *discovered a pattern of living which fitted their peculiar natural surroundings*. Societies and cultures have collapsed when they deserted their own pattern and fell into decadence, but even then, unless devastated by war, the people normally continued to provide for themselves, with

something to spare for higher things. Why not now, in so many parts of the world? I am not speaking of ordinary poverty, but of actual and acute misery; not of the poor, who, according to the universal tradition of mankind, are in a special way blessed, but of the miserable and degraded ones who, by the same tradition, should not exist at all and should be helped by all. Poverty may have been the rule in the past, but misery was not. Poor peasants and artisans have existed from time immemorial; but miserable and destitute villagers in their thousands and urban pavement dwellers in their hundreds of thousands— not in wartime or as an aftermath of war, but in the midst of peace and as a seemingly permanent feature —that is a monstrous and scandalous thing which is altogether abnormal in the history of mankind. We cannot be satisfied with the snap answer that this is due to population pressure.

"Since every mouth that comes into the world is also endowed with a pair of hands, population pressure could serve as an explanation only if it meant an absolute shortage of land—and although that situation may arise in the future, it decidedly has not arrived today (a few islands excepted). It cannot be argued that population increase as such must produce increasing poverty because the additional pairs of hands could not be endowed with the capital they needed to help themselves. Millions of people have started without capital and have shown that a pair of hands can provide not only the income but also the durable goods, i.e., capital, for civilized existence. So the question stands and demands an answer. What has gone wrong? Why cannot these people help themselves?"

The answer, I suggest, lies in the abandonment of their indigenous "culture of poverty," which means not only that they lost true culture, but also that their poverty, in all too many cases, has turned into misery.

THE COST OF THE EPHEMERAL AND THE ETERNAL

A "culture of poverty" such as we have known in innumerable variants before the industrial age is based on one fundamental distinction—which may have been made consciously or instinctively, it does not matter—the distinction between the "ephemeral" and the "eternal." All religions, of course, deal with

---

**cf. E. F. Schumacher, "Roots of Economic Growth," Gandhian Institute of Studies, Varanasi, India, 1962, pp. 37-38.

this distinction, suggesting that the ephemeral is relatively unreal and only the eternal is real. On the material plane we deal with goods and services, and the same distinction applies: all goods and services can be arranged, as it were, on a scale which extends from the ephemeral to the eternal. Needless to say, neither of these terms may be taken in an absolute sense (because there is nothing absolute on the material plane), although there may well be something absolute in the maker's *intention*: he/she may see his/her product as something to be *used up*, that is to say, to be destroyed in the act of consumption; or as something to be used or enjoyed as a permanent asset, ideally forever.

The extremes are easily recognized. An article of consumption, like a loaf of bread, is *intended* to be *used up*; while a work of art, like the Mona Lisa, is *intended* to be there forever. Transport services to take a tourist on holiday are intended to be used up and therefore ephemeral; while a bridge across the river is intended to be a permanent facility. Entertainment is intended to be ephemeral; while education (in the fullest sense) is intended to be eternal.

Between the extremes of the ephemeral and the eternal, there extends a vast range of goods and services with regard to which the producer may exercise a certain degree of choice: he/she may be producing with the intention of supplying something relatively ephemeral or something relatively eternal. A publisher, for instance, may produce a book with the intention that it should be purchased, read and treasured by countless generations; or the intention may be that it should be purchased, read and thrown away as quickly as possible.

Ephemeral goods are—to use the language of business—"depreciating assets" and have to be "written off." Eternal goods, on the other hand, are never "depreciated" but "maintained." (You don't depreciate the Taj Mahal; you try to maintain its splendour for all time.)

Ephemeral goods are subject to the economic calculus. Their only value lies in being used up, and it is necessary to ensure that their *cost of production* does not exceed the *benefit* derived from destroying them. But eternal goods are not intended for destruction: so there is no occasion for an economic calculus, because the benefit—the product of annual value and time—is infinite and therefore incalculable.

Once we recognize the validity of the distinction between the ephemeral and the eternal, we are able to distinguish, in principle, between two different types of "standard of living." Two societies may have the same volume of production and the same *income per head of population*, but the *quality of life* or life style may show fundamental and incomparable differences: the one placing its main emphasis on ephemeral satisfactions and the other devoting itself primarily to the creation of eternal values. In the former there may be opulent living in terms of ephemeral goods and starvation in terms of eternal goods—eating, drinking and wallowing in entertainment, in sordid, ugly, mean and unhealthy surroundings—while in the latter, there may be frugal living in terms of ephemeral goods and opulence in terms of eternal goods—modest, simple and healthy consumption in a noble setting. In terms of conventional economic accounting, they are both equally rich, equally developed—which merely goes to show that the purely quantitative approach misses the point.

The study of these two models can surely teach us a great deal. It is clear, however, that the question: "Which of the two is better?" reaches far beyond the economic calculus, since quality cannot be calculated.

No one, I suppose, would wish to deny that the life style of modern industrial society is one that places primary emphasis on ephemeral satisfactions and is characterized by a gross neglect of eternal goods. Under certain immanent compulsions, moreover, modern industrial society is engaged in a process of what might be called "ever-increasing ephemeralization;" that is to say, goods and services which by their very nature belong to the eternal side are being produced as if their purpose were ephemeral. The economic calculus is applied everywhere, even at the cost of skimping and cheese-paring on goods which should last forever. At the same time, purely ephemeral goods are produced to standards of refinement, elaboration and luxury, as if they were meant to serve eternal purposes and to last for all time.

Nor, I suppose, would anyone wish to deny that many pre-industrial societies have been able to create superlative cultures by placing their emphasis in the exactly opposite way. The greatest part of the modern world's cultural heritage stems from these societies.

The affluent societies of today make such exorbitant demands on the world's resources, create ecological dangers of such intensity, and produce such a high level of neurosis among their populations that they cannot possibly serve as a model to be imitated by those two-thirds or three-quarters of mankind who are conventionally considered under-developed or developing. The *failure of modern affluence*—which seems obvious enough, although it is by no means freely admitted by people of a purely materialistic outlook—cannot be attributed to affluence as such, but is directly due to mistaken priorities (the cause of which cannot be discussed here): a gross over-emphasis on the ephemeral and a brutal under-evaluation of the eternal. Not surprisingly, no amount of indulgence on the ephemeral side can compensate for starvation on the eternal side.

REDUCING WANTS TO NEEDS

In the light of these considerations, it is not difficult to understand the meaning and feasibility of a culture of poverty. It would be based on the insight that the real needs of human beings are limited and must be met, but that their wants tend to be unlimited, cannot be met, and must be resisted with the utmost determination. Only by a reduction of wants to needs can resources for genuine progress be freed. The required resources cannot be found from foreign aid; they cannot be mobilized *via* the technology of the affluent society which is immensely capital-intensive and labor-saving and is dependent on an elaborate infra-structure which is itself enormously expensive. Uncritical technology transfer from the rich societies to the poor cannot but transfer into poor societies a life style which, placing primary emphasis on ephemeral satisfactions, may suit the taste of small, rich minorities, but condemns the great, poor majority to increasing misery.

The resources for genuine progress can be found only by a life style which emphasizes frugal living in terms of ephemeral goods. Only such a life style can create, maintain and develop an ever-increasing supply of eternal goods.

Frugal living in terms of ephemeral goods means a dogged adherence to simplicity, a conscious avoidance of any unnecessary elaborations, and a magnanimous rejection of luxury—puritanism, if you like—on the ephemeral side. This makes it possible to enjoy a high standard of living on the eternal side, as a compensation and reward. Luxury and refinement have their proper place and function, but only with eternal, not with ephemeral goods. That is the essence of a culture of poverty.

One further point has to be added: the ultimate resource of any society is its labor power, which is infinitely creative. When the primary emphasis is on ephemeral goods, there is an automatic preference for mass-production, and there can be no doubt that mass production is more congenial to machines than it is to people. The result is the progressive elimination of the human factor from the productive process. For a poor society, this means that its ultimate resource cannot be properly used; its creativity remains largely untapped. This is why Gandhi, with unerring instinct, insisted that "it is not mass production but only production by the masses that can do the trick." A society that places its primary emphasis on eternal goods will automatically prefer production by the masses to mass production, because such goods, intended to last, must fit the precise conditions of their place: they cannot be standardized. This brings the whole human being back into the productive process, and it then emerges that even ephemeral goods (without which human existence is obviously impossible) are far more efficient and economical when a proper "fit" has been ensured by the human factor.

All the above does not claim to be more than an assembly of a few preliminary indications. I entertain the hope that, in view of increasing threats to the very survival of culture—and even life itself—there will be an upsurge of serious study of the possibilities of a culture of poverty. We might find that we have nothing to lose and a world to gain. ∎

Together, the values presented in this second excerpt from Tom Bender's Sharing Smaller Pies *(January, 1975) form a basis for a definition of appropriate technology. Since Tom is one of the editors of* Rain Magazine, *most of his current writings appear there. The full text of* Sharing Smaller Pies *(7) is available from* Rain *(48) as well. Reprinted with permission from the author.*

NEW VALUES

By Tom Bender
From SHARING SMALLER PIES

OUR ABILITY to develop a culture that can endure beyond our own lifetimes depends upon our coming to a new understanding of what is desirable for a harmonious and sustainable relationship with the systems that support our lives.

STEWARDSHIP, not progress.

We have valued progress highly during our period of growth, as we have known that changes were unavoidable, and have needed an orientation that could help us adjust to and assist those changes. Progress assumes that the future will be better— which at the same time creates dissatisfaction with the present and tells us that NOW isn't as good. As a result, we are prompted to work harder to get what the future can offer, but lose our ability to enjoy what we now have. We also lose a sense that we ourselves, and what we have and do, are really good. We expect the rewards from what we do to come in the future rather than from the *doing* of it, and then become frustrated when most of those dreams cannot be attained. The "future" always continues to lie in the future. Progress is really a euphemism for always believing that what we value and seek today is better than what we valued before or what anyone else has ever sought or valued.

Stewardship, in contrast to progress, elicits attentive care and concern for the present—for understanding its nature and for best developing, nurturing and protecting its possibilities. Such actions unavoidably insure the best possible future as a by-product of enjoyment and satisfaction from the present.

The government of a society has a fundamental responsibility, which we have neglected, for stewardship—particularly for the biophysical systems that support our society. It is the only organ of society which can protect those systems and protect future citizens of the society from loss of their needed resources through the profiteering of present citizens. The government's fundamental obligation in this area is to prevent deterioration in the support capacities of the biophysical systems, maintain in stable and sound fashion their ongoing capabilities, and whenever possible extend those capabilities in terms of quality as well as quantity. Present and past governments, and those who have profited from their actions, must be accountable for loss to present and future citizens and to the biophysical systems themselves from their actions.

AUSTERITY, not affluence.

Austerity is a principle which does not exclude all enjoyments, only those which are distracting from or destructive of personal relatedness. It is part of a more embracing virtue—friendship or joyfulness, and arises from an awareness that things or tools can destroy rather than enhance grace and joyfulness in personal relations. Affluence, in contrast, does not discriminate between what is wise and useful and what is merely possible. Affluence demands impossible endless growth, both because those things necessary for good relations are foregone for unnecessary things, and because many of those unnecessary things act to damage or destroy the good relations that we desire.

PERMANENCE, not profit.

Profit, as a criterion of performance, must be replaced by permanence in a world where irreplaceable resources are in scarce supply, for profit always indicates their immediate use, destroying any ability of a society to sustain itself. The only way to place lighter demands on material resources is to place heavier demands on moral resources. Permanence as a judge of the desirability of actions requires first that those actions contribute to rather than lessen the continuing quality of the society. Permanence in no way excludes fair reward for one's work—but distinguishes the profit a person gains based on loss to others from profit derived from a person's work or contribution to others.

RESPONSIBILITIES, not rights.

A society—or any relationship—based on rights rather than responsibilities is possible only when the actions involved are insignificant enough not to affect others. Our present society is based upon rights rather than responsibilities, and upon competitive distrust and contractual relationships rather than upon the more complex and cooperative kinds of relationships common in other cultures. These relationships have given us the freedom to very quickly extract and use our material wealth, settle a continent, and develop the structure of cities and civilization.

Any enduring relationship, however, must balance rights with responsibilities to prevent destruction of weaker or less aggressive, yet essential, parts of relationships—whether other people, the biosphere that supports our lives, or the various parts of our own personalities.

Distrust or contractual relationships are the easiest to escape and the most expensive to maintain—requiring the development of elaborate and expensive legal and financial systems—and cannot be the dominant form of relationship in societies that do not have the surplus wealth to afford them. Moral or ethically based relationships; relationships based on cooperation, trust and love; and the relationships encompassing more than just work, family, educational, recreational or spiritual parts of our lives are more rewarding and satisfying to the people involved. They are also more stable in their contribution to society, vastly easier to maintain and

harder to disrupt. They have always been the most common kinds of relationships between people except under the extreme duress of war or growth.

PEOPLE, not professions.

Our wealth has made it possible for us to institutionalize and professionalize many of our individual responsibilities—a process which is inherently ineffective and more costly, which has proven destructive of individual competence and confidence, and which is affordable only when significant surplus of wealth is available.

We have been able to afford going to expensively trained doctors for every small health problem, rather than learning rudiments of medical skills or taking care to prevent health problems. We have been able to afford expensive police protection rather than handling our problems by ourselves or with our neighbors. We have established professional social workers, lawyers and educators—and required that everyone use their services even for things we could do ourselves and that are wastes of the time and expertise of the professionals. As the wealth that has permitted this becomes less available to us, it will become necessary to deprofessionalize and deinstitutionalize many of these services and again take primary responsibility for them ourselves.

Our institutions have contributed to isolating, buffering and protecting us from the events of our world. This has on one hand made our lives easier and more secure, and freed us from the continual testing that is part of the dynamic interaction in any natural system. It has also, by these very actions, made us feel isolated, alienated, and rightfully fearful of not being able to meet those continued tests without the aid of our cultural and technical implements.

Our lack of familiarity with all the natural processes of our world and uncertainty of our ability to interact successfully with them aided only by our own intuitive wisdom and skills has enslaved us to those implements and degraded us. We can act confidently and with intuitive rightness only when we aren't afraid. We can open ourselves to the living interaction that makes our lives rewarding only when we cease to fear what we can't affect. Fear is only unsureness of our own abilities.

We have to take responsibilities OURSELVES for our own lives, actions, health and learning. We must also take responsibility ourselves for our community and society. There is no other way to operate any aspect of our lives and society without creating dictatorial power that destroys and prevents the unfolding of human nature and that concentrates the ability to make errors without corrective input. No one else shares our perceptions and perspective on what is occurring and its rightness, wrongness or alternatives. We are the only ones who can give that perspective to the process of determining and directing the pattern of events.

Our institutions can be tools that serve us only when they arise from and sustain the abilities of individuals and remain controlled by them.

BETTERMENT, not biggerment.

Quantitative things, because of the ease of their measurement by external means, have been sought and relied upon as measures of success by our institutionally centered society. We are learning the hard lesson that quantity is no substitute for quality in our lives, that qualitative benefits cannot be externalized, and that a society that wishes betterness rather than moreness, and betterment rather than biggerment, must be organized to allow individuals the scope for determining and obtaining what they themselves consider better.

ENOUGHNESS, not moreness.

We are learning that too much of a good thing is not a good thing, and that we would often be wiser to determine what is enough rather than how much is possible. When we can learn to be satisfied with the least necessary for happiness, we can lighten our demands on ourselves, on others, and on our surroundings, and make new things possible with what we have released from our covetousness. Our consumption ethic has prevented our thinking about enoughness, in part out of fear of unemployment problems arising from reducing our demands. Employment problems are only a result of choices of energy- vs. employment-intensive production processes and arbitrary choices we have made in the patterns of distributing the wealth of our society—both of which can be modified with little fundamental difficulty. Our major goal is to be happy with the least effort—with the least production of goods and services necessary and with the greatest opportunity to employ our time and skills for good rather than for survival. The fewer our wants, the greater our freedom from having to serve them.

LOCALIZATION, not centralization.

Centralization, in all kinds of organization, is important during periods of growth when ability to marshall resources quickly and change and direct an organization is important. It is, however, an expensive and ineffective means for dealing with ongoing operations when an excess of energy to operate the system is unavailable. As effectiveness in resolving problems on the scale and location where they occur becomes more important, organization must move to more localized and less institutionalized ways of operation. Even with sufficient resources, the power concentration of centralized systems overpowers the rights of individuals, and has proved to lead to inevitable deterioration of our quality of life.

The size and centralization of many of our organizations has nothing to do with even alleged economics or benefits of scale, and actually often is associated with diseconomics of scale and deterioration of quality of services. Size breeds size, even where it is counterproductive. It is easiest for any organization to deal with others of the same scale and kind of organization, and to create pressures for other organizations to adapt their own mode of operation.

EQUITIZATION, not urbanization.

Uncontrollable urbanization has accompanied industrialization in every country where it has occurred. The roots of that urbanization, which has occurred in spite of the desires of both the people and the governments involved, has been twofold: the destruction of traditional means of livelihood by energy slaves and the market control of large corporations, and the unequal availability of employment opportunities and educational, medical and other services. Neither of these conditions is necessary. The inequity of services has resulted from conscious choices to centralize and professionalize services rather than to manage available resources in a way to ensure equal availability of services in rural as well as urban areas. The destruction of traditional patterns of livelihood has been equally based on conscious and unnecessary choices.

Equity is not only possible, but is necessary to restore choices of where and how one lives. It is necessary to restore alternatives to our unaffordably costly urban systems. It can be achieved through introduction of appropriate technology; through

control of organization size; by equalizing income and available wealth; by establishing equal access to learning opportunities, health care, justice and other services; and by assuring everyone the opportunity for meaningful work. It can be achieved by returning to individuals the responsibility and control of their lives, surroundings and social, economic and political systems; by ensuring freedom not to consume or depend upon any systems other than one's own abilities; and by encouraging the ownership of the tools of production by the people who do the work, thus increasing the chances of developing a balanced, affluent and stable society.

WORK, not leisure.

We have considered work to be a negative thing— that the sole function of work was to produce goods and services. To workers it has meant a loss of leisure, something to be minimized while still maintaining income. To the employer it is simply a cost of production, also to be minimized. Yet work is one of our greatest opportunities to contribute to the well-being of ourselves and our community— opportunity to utilize and develop our skills and abilities, opportunity to overcome our self-centeredness through joining with other people in common tasks, as well as opportunity to produce the goods and services needed for a dignified existence. Properly appreciated, work stands in the same relation to the higher faculties as food to the physical body. It nourishes and enlivens us and urges us to produce the best of which we are capable. It furnishes a medium through which to display our scale of values and develop our personality. To strive for leisure rather than work denies that work and leisure are complementary parts of the same living process and cannot be separated without destroying the joy of work and the bliss of leisure.

From this viewpoint, work is something essential to our well-being—something that can and ought to be meaningful, the organization of which in ways which are boring, stultifying or nerve-wracking is criminal. Opportunity for meaningful work rather than merely a share of the products of work, needs to be assured to every member of our society.

TOOLS, not machines.

We need to regain the ability to distinguish between technologies which aid and those which destroy our ability to seek the ends we wish. We need to discriminate between what are tools and what are machines. The choice of tools and what they do is at root both philosophical and spiritual. Every technology has its own nature and its own effect upon the world around it. Each arises from, and supports a particular view of our world.

A tool channels work and experiences through our faculties, allowing us to bring to bear upon them the full play of our nature—to learn from the work and to infuse it with our purposes and our dreams—and to give the fullest possible opportunity for our physical and mental faculties to experience, experiment and grow. A tool focuses work so that our energy and attention can be fully employed to our chosen purposes.

Our culture has valued devices that are labor saving and require little skill to operate. By those very measures, such devices are machines which rob us of our opportunity to act, experience and grow, and to fill our surroundings with the measure of our growth. We need skill-developing rather than labor-saving technologies.

INDEPENDENCE AND INTERDEPENDENCE

Many of the basic values upon which we have tried to build our society have become weakened through the ways they have been interpreted and face the prospect of further weakening through the pressures inevitable in adapting our society to new conditions.

Independence cannot be maintained when we are dependent upon other people or other nations—as long as we are forced to work on others' terms, to consume certain kinds of education to qualify for work, to use automobiles because that kind of transportation system has made even walking dangerous or physically impossible; as long as we are dependent upon fossil fuels to operate our society; as long as we must depend upon resources other than ourselves and the renewable resources of our surroundings, we cannot be independent.

We have also discovered through the power that our wealth has given us that slavery is as enslaving for the master as for the mastered—by becoming DEPENDENT upon the abilities of the slave, whether the slave is a human, animal, institutional or energy slave, we forego developing our own capabilities to be self-sufficient.

In another sense total independence is never possible, for that means total power, which inevitably collides with the wants and power of others. We are also, in reality, dependent upon the natural systems that convert the sun's energy into the food upon which we live. Totally independent individuals may have freedom from organization, but have no special value, no special mission, no special contribution and no necessary role in the energy flows and relationships of a society that permits greater things than are attainable as individuals. Such freedom results in little respect or value for the individual. Our success and survival on this planet also must recognize the total interdependence that exists between us and the health, disease, wealth, happiness, anger and frustrations of the others with whom we share this planet.

Two things are important. We must have the CAPABILITY for self-sufficiency—in order to have options, alternatives, self-confidence and knowledge of how things are related and work and to be able to lighten our demands on others. We must also have the ABILITY to contribute our special skills to the development of interdependent relationships which can benefit all. Trade, as *giving* of *surplus*, of what is not necessary, is the only viable resolution of the interrelated problems of independence, interdependence and slavery.

As we begin to actually make changes, the things we come to find of value are almost the opposite of what we value today. What contributes to stability and soundness and to valued relationships is exactly what prevents and hinders disruption, change and growth—which have been both necessary and desired under the conditions we have until recently experienced. Meaningful work, localized economies, diversity and richness of employment and community, and controllable, clever, human-centered technologies will become important. Common sense and intuition will be recognized again as more valuable than armies of computers. Community will become more important than individualism and our present actions seen as unsupportably selfish. Strong roots and relationships will become more important than mobility. Buildings and equipment with long life and lower total costs rather than low initial costs will be favored. Cooperation will be seen as more positive, wiser and less costly than competition. Skill-using will replace labor saving. We will soon discover that all our present sciences and principles are not unbiased, but are built upon values promoting growth rather than stability, and will need to be modified when quantitative growth is no longer possible. ∎

TECHNOLOGY FOR WHAT & FOR WHOM?

Technology, like economics, mirrors values. Our choice of tools arises from the goals and aspirations of our society. Yet the results of the technologies have not always been as beneficial as we had hoped or as their proponents would have us believe: Few people have looked deeply enough into who was developing what technologies or why, and what their broader impact on our environment or our society might be. Processes that were meant to be labor-saving have proven to be dangerously mind-dulling, new drugs that were sold as miraculous cures have caused sickness; tools we idealized as creating independence have instead enforced dependence. All our so-called technological advances need re-evaluation in light of some basic questions: Who has all this progress been for? What are the costs and who pays them? Who reaps the benefits?

Some costs are subtle but pervasive. The automobile and freeway system were originally designed to move us from one place to another more quickly. Yet both have resulted in a separation of our homes from our workplaces in a way that profoundly alters the shape of our cities and the texture of our culture. Driving to work generates a considerable portion of our energy consumption used for transportation and also causes massive duplication of water, sewage, police, power and fire systems for suburbs and cities.

In addition, as Ivan Illich observes in this excerpt from Tools for Conviviality *(27), automobiles have become virtually the only method of getting around, to the exclusion of simpler, more efficient methods like bicycles, buses or walking. They, as well as schools, medical support systems and utilities, have become "radical monopolies" whose very existence forecloses other options.*

In most cases these are radical monopolies not because of natural selection or superiority of a technology, but by intent. Illich does not discuss the degree to which private corporations making and selling these technologies use their size and political means to enhance the conditions in which their products flourish. Before World War II, General Motors launched a successful campaign to eliminate the then competitive streetcar system in a number of U.S. cities—most notably Los Angeles. Los Angeles then developed its now-famous freeway system, which resulted in its equally famous pollution. And GM sold plenty of cars.

The effects of vested interests in shaping agriculture and the food production system is no more clearly documented than in the work of Frances Moore Lappé and Joseph Collins in their book, Food First *(32). The main thrust of their work is that the dominance and self-serving ends of multinational corporations and wealthy countries are more responsible than any drought for world hunger. As they document in the following excerpts, technological advances and agribusiness know-how are exclusively designed to increase corporate profits at the expense of both small farmers and the hungry people of the world.*

The infamous Green Revolution has come close to becoming a third world radical monopoly as "new improved" seeds hook farmers on a cycle of dependence on both seeds and fertilizers. Farmers were convinced to abandon hardy native seed stocks which evolved through generations of careful selection and breeding, clearly destroying individual and regional self-sufficiency while creating dependence on an exploitive few. And, as this piece shows, the new technologies aren't always what they are billed as—sometimes they don't even work.

What can be done about the apparently monolithic institutions which control our lives? Fritz Schumacher points to the direction from which change must come, saying that each solution is personal and demands a moral decision to change and create a counterbalancing force. The answer he suggests, which is the central point of this book, is technology with a human *face.*□

In Tools for Conviviality *(27), from which this piece is excerpted, Ivan Illich discusses more about the limits placed on our society by the tools we employ and the institutions we support. He further explores the pervasive effects of the automobile in* Energy and Equity—*first published in 1974 and now a chapter in his latest book,* Toward a History of Needs *(28). Reprinted by permission from Harper and Row.*

RADICAL MONOPOLY

By Ivan Illich
FROM TOOLS FOR CONVIVIALITY

WHEN OVEREFFICIENT TOOLS are applied to facilitate man's relations with the physical environment, they can destroy the balance between man and nature. Overefficient tools corrupt the environment. But tools can also be made overefficient in quite a different way. They can upset the relationship between what people need to do by themselves and what they need to obtain ready-made. In this second dimension overefficient production results in radical monopoly.

By radical monopoly I mean a kind of dominance by one product that goes far beyond what the concept of monopoly usually implies. Generally we mean by "monopoly" the exclusive control by one corporation over the means of producing (or selling) a commodity or service. Coca-Cola can create a monopoly over the soft drink market in Nicaragua by being the only maker of soft drinks which advertises with modern means. Nestlé might impose its brand of cocoa by controlling the raw material, some car maker by restricting imports of other makes, a television channel by licensing. Monopolies of this kind have been recognized for a century as dangerous by-products of industrial expansion, and legal devices have been developed in a largely futile attempt to control them. Monopolies of this kind restrict the choices open to the consumer. They might even compel him to buy one product on the market, but they seldom simultaneously abridge his liberties in other domains. A thirsty man might desire a cold, gaseous and sweet drink and find himself restricted to the choice of just one brand. He still remains free to quench his thirst with beer or water. Only if and when his thirst is translated without meaningful alternatives into the need for a Coke would the monopoly become radical. *By "radical monopoly" I mean the dominance of one type of product rather than the dominance of one brand. I speak about radical monopoly when one industrial production process exercises an exclusive control over the satisfaction of a pressing need, and excludes nonindustrial activities from competition.* [Emphasis added.]

Cars can thus monopolize traffic. They can shape a city into their image—practically ruling out locomotion on foot or by bicycle in Los Angeles. They can eliminate river traffic in Thailand. That motor traffic curtails the right to walk, not that more people drive Chevies than Fords, constitutes radical monopoly. What cars do to people by virtue of this radical monopoly is quite distinct from and independent of what they do by burning gasoline that

could be transformed into food in a crowded world. It is also distinct from automotive manslaughter. Of course cars burn gasoline that could be used to make food. Of course they are dangerous and costly. But the radical monopoly cars establish is destructive in a special way. Cars create distance. Speedy vehicles of all kinds render space scarce. They drive wedges of highways into populated areas, and then extort tolls on the bridge over the remoteness between people that was manufactured for their sake. This monopoly over land turns space into car fodder. It destroys the environment for feet and bicycles. Even if planes and buses could run as nonpolluting, nondepleting public services, their inhuman velocities would degrade man's innate mobility and force him to spend more time for the sake of travel.

Schools tried to extend a radical monopoly on learning by redefining it as education. As long as people accepted the teacher's definition of reality, those who learned outside school were officially stamped "uneducated." Modern medicine deprives the ailing of care not prescribed by doctors. Radical monopoly exists where a major tool rules out natural competence. Radical monopoly imposes compulsory consumption and thereby restricts personal autonomy. It constitutes a special kind of social control because it is enforced by means of the imposed consumption of a standard product that only large institutions can provide.

The control of undertakers over burial shows how radical monopoly functions and how it differs from other forms of culturally defined behavior. A generation ago, in Mexico, only the opening of the grave and the blessing of the dead body were performed by professionals: the gravedigger and the priest. A death in the family created various demands, all of which could be taken care of within the family. The wake, the funeral, and the dinner served to compose quarrels, to vent grief, and to remind each participant of the fatality of death and the value of life. Most of these were of a ritual nature and carefully prescribed—different from region to region. Recently, funeral homes were established in the major cities. At first undertakers had difficulty finding clients because even in large cities people still knew how to bury their dead. During the sixties the funeral homes obtained control over new cemeteries and began offering package deals, including the casket, church service and embalming. Now legislation is being passed to make the mortician's ministrations compulsory. Once he gets hold of the body, the funeral director will have established a radical monopoly over burial, as medicine is at the point of establishing one over dying.

The current debate over health-care delivery in the United States clearly illustrates the entrenchment of a radical monopoly. Each political party in the debate makes sick-care a burning public issue and thereby relegates health care to an area about which politics has nothing important to say. Each party promises more funds to doctors, hospitals and drugstores. Such promises are not in the interest of the majority. They only serve to increase the power of a minority of professionals to prescribe the tools men are to use in maintaining health, healing sickness and repressing death. More funds will strengthen the hold of the health industry over public resources and heighten its prestige and arbitrary power. Such power in the hands of a minority will produce only an increase in suffering and a decrease in personal self-reliance. More money will be invested in tools that only postpone unavoidable death and in services that abridge even further the civil rights of those who want to heal each other. More money spent under the control of the health profession means that more people are operationally conditioned into playing the role of the sick, a role they are not allowed to interpret for themselves. Once they accept this role, their most trivial needs can be satisfied only through commodities that are scarce by professional definition.

People have a native capacity for healing, consoling, moving, learning, building their houses and burying their dead. Each of these capacities meets a need. The means for the satisfaction of these needs are abundant so long as they depend pri-

marily on what people can do for themselves, with only marginal dependence on commodities. These activities have use-value without having been given exchange-value. Their exercise at the service of man is not considered labor.

These basic satisfactions become scarce when the social environment is transformed in such a manner that basic needs can no longer be met by abundant competence. The establishment of radical monopoly happens when people give up their native ability to do what they can do for themselves and for each other in exchange for something "better" that can be done for them only by a major tool. Radical monopoly reflects the industrial institutionalization of values. It substitutes the standard package for the personal response. It introduces new classes of scarcity and a new device to classify people according to the level of their consumption. This redefinition raises the unit cost of valuable service, differentially rations privilege, restricts access to resources, and makes people dependent. Above all, by depriving people of the ability to satisfy personal needs in a personal manner, radical monopoly creates radical scarcity of personal—as opposed to institutional—service.

Against this radical monopoly people need protection. They need this protection whether consumption is imposed by the private interests of undertakers, by the government for the sake of hygiene, or by the self-destructive collusion between the mortician and the survivors, who want to do the best thing for their dear departed. They need this protection even if the majority is now sold on the professional's services. Unless the need for protection from radical monopoly is recognized, its multiple implementation can break the tolerance of man for enforced inactivity and passivity.

It is not always easy to determine what constitutes compulsory consumption. The monopoly held by schools is not established primarily by a law that threatens punishment to parent or child for truancy. Such laws exist, but school is established by other tactics: by discrimination against the unschooled, by centralizing learning tools under the control of teachers, by restricting public funds earmarked for babysitting to salaries for graduates from normal schools. Protection against laws that impose education, vaccination or life prolongation is important, but it is not sufficient. Procedures must be used that permit any party who feels threatened by compulsory consumption to claim protection, whatever form the imposition takes. Like intolerable pollution, intolerable monopoly cannot be defined in advance. The threat can be anticipated, but the definition of its precise nature can result only from people's participation in deciding what may not be produced.

Protection against this general monopoly is as difficult as protection against pollution. People will face a danger that threatens their own self-interest but not one that threatens society as a whole. Many more people are against cars than are against driving them. They are against cars because they pollute and because they monopolize traffic. They drive cars because they consider the pollution created by one car insignificant, and because they do not feel personally deprived of freedom when they drive. It is also difficult to be protected against monopoly when a society is already littered with roads, schools or hospitals, when independent action has been paralyzed for so long that the ability for it seems to have atrophied, and when simple alternatives seem beyond the reach of the imagination. Monopoly is hard to get rid of when it has frozen not only the shape of the physical world but also the range of behavior and of imagination. Radical monopoly is generally discovered only when it is too late.

Commercial monopoly is broken at the cost of the few who profit from it. Usually, these few manage to evade controls. The cost of radical monopoly is already borne by the public and will be broken only if the public realizes that it would be better off paying the costs of ending the monopoly than by continuing to pay for its maintenance. But the price will not be paid unless the public learns to value the potential of a convivial society over the illusion of progress. It will not be paid voluntarily by those who confuse conviviality with intolerable poverty.

Some of the symptoms of radical monopoly are reaching public awareness, above all the degree to which frustration grows faster than output in even the most highly developed countries and under whatever political regime. Policies aimed to ease this frustration may easily distract attention from the general nature of the monopoly at its roots, however. The more these reforms succeed in correcting superficial abuses, the better they serve to bolster the monopoly I am trying to describe.

The first palliative is consumer protection. Consumers cannot do without cars. They buy different makes. They discover that most cars are unsafe at any speed. So they organize to get safer, better and more durable cars and to get more as well as wider and safer roads. Yet when consumers gain more confidence in cars, the victory only increases society's dependence on high-powered vehicles—public or private—and frustrates even more those who have to, or would prefer to, walk.

While the organized self-protection of the addict-consumer immediately raises the quality of the dope and the power of the peddler, it also may lead ultimately to limits on growth. Cars may finally become too expensive to purchase and medicines too expensive to test. By exacerbating the contradictions inherent in this institutionalization of values, majorities can more easily become aware of them. Discerning consumers who are discriminatory in their purchasing habits may finally discover that they can do better by doing things for themselves.

The second palliative proposed to cure growing frustration with growing output is planning. The illusion is common that planners with socialist ideals might somehow create a socialist society in which industrial workers constitute a majority. The proponents of this idea overlook the fact that anticonvivial and manipulative tools can fit into a socialist society in only a very limited measure. Once transportation, education or medicine is offered by a government free of cost, its use can be enforced by moral guardians. The underconsumer can be blamed for sabotage of the national effort. In a market economy, someone who wants to cure his flu by staying in bed will be penalized only through loss of income. In a society that appeals to the "people" to meet centrally determined production goals, resistance to the consumption of medicine becomes an act of public immorality. Protection against radical monopoly depends on a political consensus opposed to growth. Such a consensus is diametrically opposed to the issues now raised by political oppositions, since these converge in the demand to increase growth and to provide more and better things for more completely disabled people.

Both the balance that defines man's need for a hospitable environment and the balance that defines everyone's need for authentic activity are now close to the breaking point. And still this danger does not concern most people. It must now be explained why most people are either blind to this threat or feel helpless to correct it. I believe that the blindness is due to the decline in a third balance—the balance of learning—and that the impotence people experience is the result of yet a fourth upset in what I call the balance of power. ∎

Food First *(32)* *is an important book for anyone who wants to dig more deeply into corporate influence and the world food situation. It is available in hardback from the Institute for Food Development Policy* (32) *and will soon be out in paperback. Two other important related books which touch both the personal and political aspects of food and corporate control are* Diet for a Small Planet *(31) by Frances Moore Lappé and* Global Reach *(5) by Richard Barnet and Ronald Muller. Reprinted by permission from the Institute for Food Development Policy.*

ISN'T NATURE NEUTRAL?

By Frances Moore Lappé and Joseph Collins
From FOOD FIRST

As PART OF A FIFTEEN-NATION STUDY of the impact of the new seeds* conducted by the United Nations Research Institute for Social Development, Dr. Ingrid Palmer concludes that the term "high-yielding varieties" (HYVs) is a misnomer because it implies that the new seeds are high-yielding *in and of themselves*.[1] The distinguishing feature of the seeds, however, is that they are highly *responsive* to certain key inputs such as irrigation and fertilizer. Following Palmer's lead we have chosen to use the term "high-response varieties" (HRVs) as much more revealing of the true character of the seeds. The Green Revolution is obviously more complicated than just sticking new varieties of seeds into the ground. Unless the poor farmers can afford to ensure the ideal conditions that will make these new seeds respond (in which case they wouldn't be poor!), their new seeds are just not going to grow as well as the ones planted by better-off farmers. The new seeds prefer the "better neighborhoods."

Just as significant for the majority of the world's farmers is that the new seeds show a greater yield variation than the seeds they displace.[2] The HRVs are more sensitive to drought and flood than their traditional predecessors. They are particularly prone to water stress—the inability to assimilate nutrients when not enough water is getting to the plant roots, especially at certain stages in their growth cycles. Under these conditions it is often no more profitable to apply fertilizers to the new seeds than to the previous ones.[3] In 1968-1969 in Pakistan, for example, yields of Mexican dwarf wheat declined by about 20 percent because of a two-thirds reduction in average rainfall and higher than normal temperatures. The locally adapted varieties, however, were not adversely affected by the weather changes. Instead their yields increased 11 percent.[4] The new sorghums now being planted in Upper Volta in Africa are also less drought-resistant than their local cousins.[5] The HRVs can also be more vulnerable to too much water. Being shorter, the HRVs of rice cannot tolerate the higher flood levels that indigenous varieties can endure.

Since the HRVs are more sensitive to both too much and too little water, they need, not mere irrigation, but sophisticated water management. The significance of this need becomes clear in India's Punjab. Higher yields from the new seeds depend

---

*high-yielding seeds developed as part of the Green Revolution.

58

on a tubewell for a controlled water supply. But a tubewell is well beyond the means of the small farmer.

Taking advantage of the HRVs has required farmers to double or even triple their indebtedness. Since small farmers are already in debt for preharvest consumption and for other family needs—often at very high rates of interest—most will not be able to take on this heavy new burden.

HRVs are often less resistant to disease and pests. Vulnerability results from transplanting a variety that "evolved" over a short period in one climate (with a little help from agronomists) to an entirely different climate, thus supplanting varieties that had evolved over centuries in response to a long history of natural threats in that environment. A small farmer, whose family's very survival depends on each and every harvest, cannot afford to risk crop failure. For the large farmer that risk is minimized. The difference is not just that the large farmer can better withstand a crop failure. Large farmers have also managed in many instances to protect themselves against disease-prone seeds. In Mexico, for example, associations of large farmers have, since the early days of the Green Revolution, kept in constant touch with government seed agencies and

have, therefore, been warned of any disease likely to attack a particular new plant variety. By way of constrast, when the agency to which the landreform beneficiaries were tied was notified that its seeds had become susceptible to disease, it was reluctant to throw them out and often continued to sell them. To a large extent this accounts for the disastrous yields in the Mexican land-reform sector during the late 1950s and early 1960s.[6]

In addition, the new seeds have been restricted to well rainfed and irrigated regions. It is not coincidental that these favored regions are inhabited by the more affluent farmers. Almost all of the HRV increases in wheat cultivation in India have taken place in the states of Punjab and Haryana, largely because the soil is alluvial and a canal system assures a year-round water supply.[7]

Nyle C. Brady, Director of the International Rice Research Institute, where many of the new strains have been developed, estimates that the "new rice varieties may be suitable for only 25 percent of the world's acreage, largely those areas with water for irrigation."[8] Because the new varieties are less resistant to flooding, there are many parts of Thailand, Bangladesh and South Vietnam in which they cannot be used.[9] None of the new seeds are successful in areas of constant high temperatures and rainfall, limited sun and thin, badly leached soils.

Knowing the biological requirements of the seeds, we should not be surprised that as late as 1972-1973 the HRVs covered a very small percentage—only about *15* percent of the total world area excluding the socialist countries.[10] Furthermore, they are highly concentrated: 81 percent of the HRV wheat grows in a small area in India and in Pakistan; and four countries (India, the Philippines, Indonesia and Bangladesh) account for 83 percent of the HRV rice.[11]

The seeds, due to their need for ideal conditions, are restricted to certain favored areas. They therefore have reinforced income disparities between geographic regions, just as they have exacerbated the inequalities between social classes.

Two other factors contribute to making the new seeds less than neutral. First, hybrids of corn and sorghum do not remain genetically pure year after year. To maintain high yields new hybrid seeds must be purchased each year. This requirement alone gives the edge to the wealthier farmer and to the farmer more closely linked to seed distributors and other credit sources. The many farmers with only enough land to grow the food their families need will never have the cash to purchase hybrid seeds.

Second, the new seeds, because they require special knowledge to be used effectively, are inherently biased in favor of those who have access to government agricultural extension agents and instruction literature. In many countries the large landowners have been able to monopolize the services of the extension agencies. They have also been able, as in the Mexican state of Sonora, to hire private agricultural and pest experts to solve their technical problems.[12] A study in Uttar Pradesh, India, showed that, since 70 percent of the family heads were completely illiterate, "access to literature is thus primarily the prerogative of the better educated, wealthier landowners." Nonwritten material was no more successful in overcoming the problem: the village headman regarded the radio as his private property and invited only his friends to listen.[13]

The bias can be quite subtle. As one student of the Green Revolution so aptly put it: "The new technology puts a relative handicap on those whose assets include traditional knowledge of the local idiosyncrasies of soil and climate and whose energies are absorbed by the labors of husbandry. . . . It gives the advantage to those skilled in manipulating influence."[14]

Still the idea that a seed, the product of impartial scientific research, must be neutral, without built-in bias, is deeply rooted in most of us. Most assume it will just be a matter of time before the new seeds spread out to the poor and bring the standard of living up for all farmers. But the dependence of HRVs on *optimal* conditions make that impossible in most areas today. Both the rich and the poor farmer certainly can plant the seed, but who can feed the plants the optimal diet of nutrients and water and protect them from disease and pests? Can the family who depends for their food on what they grow afford to gamble with the less dependable seeds?

The only way that such seeds can be neutral is if the society prepares the way—giving equal access to the necessary inputs to all farmers. If this means redistribution of control over all food-producing resources, including land redistribution, it can work. In Cuba, for example, between 75 and 90 percent of the rice acreage is planted with the high-response seeds.[15] In Taiwan, also a country of fairly equal land distribution, the use of improved seeds is over 90 percent. But where "equalizing access" has merely meant credit programs it has rarely worked.

Any notion of equalizing access to a new technology without altering the basic social structure overlooks the only truly workable approach—the farmers themselves becoming the innovators. Then the issue of dissemination of new seeds or new skills evaporates as a problem. And, as you might guess, the kinds of seeds developed are not those demanding ideal conditions.

In China, production of the new seeds does not take place in central experimental stations but is handled by ordinary families themselves. Most communes have their own laboratories for locally developing new varieties. Spreading the new technology is therefore not a problem. As early as 1961, the Chinese were breeding seeds for less favorable climes. Chinese farmers have successfully developed seeds that are both higher yielding and *more* able to withstand bad weather and other dangers, such as barley strains adapted to high altitudes and cold-resistant strains of wheat.

Once manipulated by people, nature loses its neutrality. Elite research institutes will produce new seeds that work—at least in the short-term—for a privileged class of commercial farmers. Genetic research that involves ordinary farmers themselves will produce seeds that are useful to them. A new seed, then, is like any other technological development; its contribution to social progress depends entirely on who develops it and who controls it.

• • •

TRANSFER OF TECHNOLOGY OR TRANSFER OF ASSUMPTIONS?

To most Americans the release of human beings from labor in the fields is equated with efficiency and progress. (How could anyone but the most "pathologically anti-technological" person see it any other way?)[20] Who would not prefer to help people escape a backbreaking life in the fields for an easier life of the city? In most cases, however, the introduction of large-scale, high-energy agricultural technology does not suddenly give the agricultural workers the option to leave. It gives them rather no other otpion *but* to leave. Indeed, the same may be said of small farmers in America today. We cannot look at technology in agriculture as freeing people from labor when productive work is what people need and want more than anything else.

Although large-scale mechanization is not necessary for increased production, Americans have a hard time believing it. Why? In part because we tend to think that machines are necessary for doing all the big jobs in life. Our notions are based on our own self-conceptions. Most of us accomplish very little with our bodies. Maybe our arms and hands shift a gear; our fingers push a button. But we know little of what our bodies are capable of creating. Naturally, we come to assume big machines are necessary.

A most impressive and visible example that counters this assumption is the transformation of eroded valleys in China into a series of fertile, terraced fields—almost all accomplished *without* machinery. H. V. Henley's FAO study of China recounts the logical way the Chinese people put gravity to work for them. Large dams were built mainly by hand in the eroded and barren valleys. Slowly they silted up to what eventually became a new terraced field, then another dam was built a little higher in the valley. Gradually, a series of terraced steps was created.[21]

Perhaps another reason Americans tend to assume that large machines are essential is that we live in a society where we are made to feel isolated from and competitive with our neighbors. The po-

tential of collective action, therefore, is seldom even considered and virtually never experienced. ∎

1 Ingrid Palmer, *Science and Agricultural Production* (Geneva: UNRISD, 1972), pp. 6-7.
2 World Bank, *The Assault on World Poverty–Problems of Rural Development, Education, and Health* (Baltimore: Johns Hopkins University Press, 1975), pp. 132-133.
3 Andrew Pearse, "Social and Economic Implications of the Large-Scale Introduction of the New Varieties of Foodgrains, Part 2 (Geneva: UNDP/UNRISD, 1975), p. II-7.
4 S. Ahmed and S. Abu Khalid, "Why Did Mexican Dwarf Wheat Decline in Pakistan? *World Crops* 23:211-215.
5 Charles Elliott, *Patterns of Poverty in the Third World—A Study of Social and Economic Stratification* (New York: Praeger, 1975), pp. 47-48.
6 Cynthia Hewitt de Alcantara, "The Social and Economic Implications of the Large-Scale Introduction of the New Varieties of Foodgrains," *Country Report—Mexico* (Geneva: UNDP/UNRISD, 1974), p. 181.
7 North London Haslemere, *The Death of the Green Revolution* (London: Haslemere Declaration Group; Oxford: Third World First), p. 4.
8 Victor McElheny, "Nations Demand Agricultural Aid," *New York Times*, Aug. 3, 1975, p. 20.
9 Keith Griffin, *The Political Economy of Agrarian Change* (Cambridge, Mass.: Harvard University Press, 1974), p. 205.
10 Pearse, "Social and Economic Implications," Part 1, pp. 111-118.
11 Nicholas Wade, "Green Revolution I: A Just Technology Often Unjust in Use," *Science* (Dec. 1974): 1093-1096.
12 Hewitt de Alcántara, *Country Report—Mexico*, p. 87.
13 Pearse, "Social and Economic Implications," Part 4, pp. XI-52, XI-53.
14 Pearse, "Social and Economic Implications," Part 3, pp. IX-23, IX-24.
15 Palmer, *Science and Agricultural Production*, p. 47.
16 Food and Agricultural Organization, *Report on China's Agriculture*, prepared by H. V. Henle, 1974, pp. 144-145.
17 Erich M. Jacoby, *The "Green Revolution" in China* (Geneva: UNRISD, 1974), p. 6.

• • •

20 Ray Vicker, *This Hungry World* (New York: Scribner's, 1975)
21 Food and Agriculture Organization, *Report on China's Agriculture*, prepared by H. V. Henle, 1974, p. 148.

Railroad surveys must be received with great
caution where any motive exists for *cooking* them.
Capitalists are shy of investments in roads with step
grades, and of course it is important to make a fair
show of facilities in obtaining funds for new routes.
Joint-stock companies have no souls; their managers, in general, no consciences. Cases can be cited
where engineers and directors of railroads, with
long grades above 100 feet to the mile, have regularly sworn in their annual reports, for years in succession, that there were no grades upon their routes
exceeding half that elevation. In fact, every person
conversant with the history of these enterprises
knows that in their public statements falsehood is
the rule, truth the exception.

What I am about to remark is not exactly relevant
to my subject; but it is hard to "get the floor" in the
world's great debating society, and when a speaker
who has anything to say once finds access to the
public ear, he must make the most of his opportunity, without inquiring too nicely whether his observations are "in order." I shall harm no honest man
by endeavoring, as I have often done elsewhere, to
excite the attention of thinking and conscientious
men to the dangers which threaten the great moral
and even political interests of Christendom, from
the unscrupulousness of the private associations
that now control the monetary affairs, and regulate
the transit of persons and property, in almost every
civilized country. More than one American State is
literally governed by unprincipled corporations,
which not only defy the legislative power, but have,
too often, corrupted even the administration of justice. Similar evils have become almost equally rife in
England, and on the Continent; and I believe the
decay of commercial morality, and I fear of the sense
of all higher obligations than those of a pecuniary
nature, on both sides of the Atlantic, is to be
ascribed more to the influence of joint-stock banks
and manufacturing and railway companies, to the
workings, in short, of what is called the principle of
"associate action," than to any other one cause of
demoralization.

G. P. Marsh
From MAN AND NATURE, 1894 (39)

TIME TO STOP

By E. F. Schumacher
From TECHNOLOGY AND POLITICAL CHANGE

Not very long ago I visited a famous institution
developing textile machinery. The impression is
overwhelming. The latest and best machines, it
seemed to me, can do everything I could possibly
imagine; in fact, more than I could imagine before
I saw them.

"You can now do everything," I said to the professor who was taking me around; "Why don't you
stop, call it a day?"

My friendly guide did indeed stop in his tracks:
"My goodness!" he said, "what do you mean? You
can't stop progress. I have all these clever people
around me who can still think of improvements.
You don't expect me to suppress good ideas?
What's wrong with progress?"

"Only that the price per machine, which is already around the 100,000 pound mark, will rise to
150,000 pounds."

"But what's wrong with that?" he demanded.
"The machine will be 50 percent dearer but at least
60 percent better."

"Maybe," I replied, "but also that much more
exclusive to the rich and powerful. Have you ever
reflected on the *political* effect of what you are
doing?"

Of course, he had never given it a thought. But he
was much disturbed; he saw the point at once. "I
can't stop," he pleaded.

"Of course, you can't stop. But you can do some-

This piece is excerpted from a longer essay entitled Technology and Political Change *(53), which was first printed in* Resurgence *(50) in 1976. It is available in full from them or from* Rain Magazine *(48) in the December, 1976, and January, 1977, issues (Vol. II, Nos. 3 and 4) for $1 each. Reprinted with permission from* Resurgence.

thing all the same: you can strive to create a counter-weight, a counterforce, namely, efficient small-scale technology for the little people. What are you in fact doing for the little people?"

"Nothing."

I talked to him about what I call the "Law of the Disappearing Middle." In technological development, when it is drifting along, outside conscious control, all ambition and creative talent *goes to the frontier*, the only place considered prestigious and exciting.

Development proceeds from Stage 1 to Stage 2, and when it moves on to Stage 3, Stage 2 drops out; when it moves on to Stage 4, Stage 3 drops out, and so on.

It is not difficult to observe the process. The "better" is the enemy of the good and makes the good disappear *even if* most people cannot afford the better, for reasons of Money, Market, Management, or whatever it might be. Those who cannot afford to keep pace drop out and are left with nothing but Stage 1 technology. If, as a farmer, you cannot afford a tractor and a combine harvester, where can you get efficient animal-drawn equipment for these jobs—the kind of equipment I myself used 35 years ago? Hardly anywhere. *So you cannot stay in farming*. The hoe and the sickle remain readily available; the latest and the best—for those who can afford it—is also readily available. But the middle, the intermediate technology, disappears. Where it does not disappear altogether it suffers from total neglect—no improvements, no benefits from any new knowledge, antiquated, unattractive, etc.

The result of all this is a loss of freedom. The power of the rich and powerful becomes ever more all-embracing and systematic. The free and independent "middle class," capable of challenging the monopolistic power of the rich, disappears in step with the "disappearing middle" of technology. (There remains a middle class of managerial and professional servants of the rich organizations; they cannot challenge anything.) Production and incomes become concentrated in fewer and fewer hands, or organizations, or bureaucracies—a tendency which redistributive taxation plus ever-increasing welfare payments frantically try to counteract—and the rest of mankind have to hawk themselves around to find a "slot" provided by the rich, into which they might fit. The First Commandment is: Thou shalt adapt thyself. To what? To the available "slots." And if there are not enough of them available, you are left unemployed. Never previously having done your own thing, it is unlikely that you will have the ability to do it now, and in any case the technology that could help you to do your own thing *efficiently* cannot be found. ■

Written in 1973, this piece is excerpted from the chapter called "Technology with a Human Face" in Small Is Beautiful *(52). Even though it focuses largely on the Third World, this best seller is required reading for anyone wanting to develop a complete understanding of the problems inherent in large-scale, capital-intensive technologies and the possibilities of those more humanly proportioned. Reprinted with permission from Harper and Row.*

TECHNOLOGY WITH A HUMAN FACE

By E. F. Schumacher
From SMALL IS BEAUTIFUL

THE MODERN WORLD has been shaped by its metaphysics, which has shaped its education, which in turn has brought forth its science and technology. So, without going back to metaphysics and education, we can say that the modern world has been shaped by technology. It tumbles from crisis to crisis; on all sides there are prophecies of disaster and, indeed, visible signs of breakdown.

If that which has been shaped by technology, and continues to be so shaped, looks sick, it might be wise to have a look at technology itself. If technology is felt to be becoming more and more inhuman, we might do well to consider whether it is possible to have something better—a technology with a human face.

Strange to say, technology, although of course the product of man, tends to develop by its own laws and principles, and these are very different from those of human nature or of living nature in general. Nature always, so to speak, knows where and when to stop. Greater even than the mystery of natural growth is the mystery of the natural cessation of growth. There is measure in all natural things—in their size, speed, or violence. As a result, the system of nature, of which man is a part, tends to be self-balancing, self-adjusting, self-cleansing. Not so with technology, or perhaps I should say: not so with man dominated by technology and specialization. Technology recognizes no self-limiting principle—in terms, for instance, of size, speed, or violence. It therefore does not possess the virtues of being self-balancing, self-adjusting and self-cleansing. In the subtle system of nature, technology, and in particular the super-technology of the modern world, acts like a foreign body, and there are now numerous signs of rejection.

Suddenly, if not altogether surprisingly, the modern world, shaped by modern technology, finds itself involved in three crises simultaneously. First, human nature revolts against inhuman technological, organizational, and political patterns, which it experiences as suffocating and debilitating; second, the living environment which supports human life aches and groans and gives signs of partial breakdown; and third, it is clear to anyone fully knowledgeable in the subject matter that the inroads being made into the world's non-renewable resources, particularly those of fossil fuels, are such that serious bottlenecks and virtual exhaustion loom ahead in the quite forseeable future.

Any one of these three crises or illnesses can turn out to be deadly. I do not know which of the three is the most likely to be the direct cause of collapse. What is quite clear is that a way of life that bases itself on materialism, *i.e.* on permanent, limitless expansionism in a finite environment, cannot last long, and that its life expectation is the shorter the more successfully it pursues its expansionist objectives.

If we ask where the tempestuous developments of world industry during the last quarter-century have taken us, the answer is somewhat discouraging. Everywhere the problems seem to be growing faster than the solutions. This seems to apply to the rich countries just as much as to the poor. There is nothing in the experience of the last 25 years to suggest that modern technology, as we know it, can really help us to alleviate world poverty, not to mention the problem of unemployment which already reaches levels like 30 percent in many so-called developing countries, and now threatens to become endemic also in many of the rich countries.

In any case, the apparent yet illusory successes of the last 25 years cannot be repeated: the threefold crisis of which I have spoken will see to that. So we had better face the question of technology—what does it do and what should it do? Can we develop a technology which really helps us to solve our problems—a technology with a human face?

The primary task of technology, it would seem, is to lighten the burden of work man has to carry in order to stay alive and develop his potential. It is easy enough to see that technology fulfills this purpose when we watch any particular piece of machinery at work—a computer, for instance, can do in seconds what it would take clerks or even mathematicians a very long time, if they can do it at all. It is more difficult to convince oneself of the truth of this simple proposition when one looks at whole societies. When I first began to travel the world, visiting rich and poor countries alike, I was tempted to formulate the first law of economics as follows: "The amount of real leisure a society enjoys tends to be in inverse proportion to the amount of labour-saving machinery it employs." It might be a good idea for the professors of economics to put this proposition into their examination papers and ask their pupils to discuss it. However that may be, the evidence is very strong indeed. If you go from easy-going England to, say, Germany or the United States, you find that people there live under much more strain than here. And if you move to a country like Burma, which is very near to the bottom of the league table of industrial progress, you find that people have an enormous amount of leisure really to enjoy themselves. Of course, as there is so much less labor-saving machinery to help them, they "accomplish" much less than we do; but that is a different point. The fact remains that the burden of living rests much more lightly on their shoulders than on ours.

The question of what technology actually does for us is therefore worthy of investigation. It obviously greatly reduces some kinds of work while it increases other kinds. The type of work which modern technology is most successful in reducing or even eliminating is skillful, productive work of human hands, in touch with real materials of one kind or another. In an advanced industrial society, such work has become exceedingly rare, and to make a decent living by doing such work has become virtually impossible. A great part of the modern neurosis may be due to this very fact; for the human being, defined by Thomas Aquinas as a being with brains and hands, enjoys nothing more than to be creatively, usefully, productively engaged with both his hands and his brains. Today, a person has to be wealthy to be able to enjoy this simple thing, this very great luxury: he has to be lucky enough to find a good teacher and plenty of free time to learn and practice. He really has to be rich enough not to need a job; for the number of jobs that would be satisfactory in these respects is very small indeed.

The extent to which modern technology has taken over the work of human hands may be illustrated as follows. We may ask how much of "total social time"—that is to say, the time all of us have together, 24 hours a day each—is actually engaged in real production. Rather less than one-half of the total population of this country is, as they say, gainfully occupied, and about one-third of these are actual producers in agriculture, mining, construction and industry. I do mean *actual producers*, not people who tell other people what to do, or account for the past, or plan for the future, or distribute what other people have produced. In other words, rather less than one-sixth of the total population is engaged in actual production; on average, each of them supports five others beside himself, of which two are gainfully employed on things other than real production and three are not gainfully employed. Now, a fully employed person, allowing for holidays, sickness, and other absence, spends about one-fifth of his total time on his job. It follows that the proportion of "total social time" spent on actual production—in the narrow sense in which I am using the term—is, roughly, one-fifth of one-third of one-half, *i.e.* 3½ percent. The other 96½ percent of "total social time" is spent in other ways, including sleeping, eating, watching television, doing jobs that are not *directly* productive, or just killing time more or less humanely.

Although this bit of figuring work need not be taken too literally, it quite adequately serves to show what technology has enabled us to do: namely, to reduce the amount of time actually spent on production in its most elementary sense to such a tiny percentage of total social time that it pales into insignificance, that it carries no real weight, let alone prestige. When you look at industrial society in this way, you cannot be surprised to find that prestige is carried by those who help fill the other 96½ percent of total social time, primarily the entertainers but also the executors of Parkinson's Law. In fact, one might put the following proposition to students of sociology: "The prestige carried by people in modern industrial society varies in inverse proportion to their closeness to actual production."

There is a further reason for this. The process of confining productive time to 3½ percent of total social time has had the inevitable effect of taking all normal human pleasure and satisfaction out of the time spent on this work. Virtually all real production has been turned into an inhuman chore which does not enrich a man but empties him. "From the factory," it has been said, "dead matter goes out improved, whereas men there are corrupted and degraded."

We may say, therefore, that modern technology has deprived man of the kind of work that he enjoys most, creative, useful work with hands and brains, and given him plenty of work of a fragmented kind, most of which he does not enjoy at all. It has multiplied the number of people who are exceedingly busy doing kinds of work which, if it is productive at all, is so only in an indirect or "roundabout" way, and much of which would not be necessary at all if technology were rather less modern. Karl Marx appears to have foreseen much of this when he wrote: "They want production to be limited to useful things, but they forget that the production of too many useful things results in too many useless people," to which we might add: particularly when the processes of production are joyless and boring. All this confirms our suspicion that modern technology, the way it has developed, is developing, and promises further to develop, is showing an increasingly inhuman face, and that we might do well to take stock and reconsider our goals.

Taking stock, we can say that we possess a vast accumulation of new knowledge, splendid scientific techniques to increase it further, and immense experience in its application. All this is truth of a kind. This truthful knowledge, as such, does *not* commit us to a technology of giantism, supersonic speed, violence and the destruction of human work-enjoyment. The use we have made of our knowledge is only one of its possible uses and, as is now becoming ever more apparent, often an unwise and destructive use.

As I have shown, directly productive time in our society has already been reduced to about 3½ percent of total social time, and the whole drift of modern technological development is to reduce it further, asymptotically* to zero. Imagine we set ourselves a goal in the opposite direction—to increase it sixfold, to about 20 percent, so that 20 percent of total social time would be used for actually producing things, employing hands and brains and, naturally, excellent tools. An incredible thought! Even children would be allowed to make themselves useful, even old people. At one-sixth of present-day productivity, we should be producing as much as at present. There would be six times as much time for any piece of work we chose to undertake—enough to make a really good job of it, to enjoy oneself, to produce real quality, even to make things beautiful. Think of the therapeutic value of real work; think of its educational value. No one would then want to raise the school-leaving age or to lower the retirement age, so as to keep people off the labor market. Everybody would be welcome to lend a hand. Everybody would be admitted to what is now the rarest privilege, the opportunity of working usefully, creatively, with his own hands and brains, in his own time, at his own pace—and with excellent tools. Would this mean an enormous extension of working hours? No, people who work in this way do not know the difference between work and leisure. Unless they sleep or eat or occasionally choose to do nothing at all, they are always agreeably, productively engaged. Many of the "on-cost jobs" would simply disappear; I leave it to the reader's imagination to identify them. There would be little need for mindless entertainment or other drugs, and unquestionably much less illness.

Now, it might be said that this is a romantic, a utopian, vision. True enough. What we have today, in modern industrial society, is not romantic and certainly not utopian, as we have it right here. But it is in very deep trouble and holds no promise of survival. We jolly well have to have the courage to dream if we want to survive and give our children a chance of survival. The threefold crisis of which I have spoken will not go away if we simply carry on as before. It will become worse and end in disaster, until or unless we develop a new life-style which is compatible with the real needs of human nature, with the health of living nature around us, and with the resource endowment of the world.

Now, this is indeed a tall order, not because a new life-style to meet these critical requirements

---

*Asymptote: A mathematical line continually approaching some curve but never meeting it within a finite distance.

67

and facts is impossible to conceive, but because the present consumer society is like a drug addict who, no matter how miserable he may feel, finds it extremely difficult to get off the hook. The problem children of the world—from this point of view and in spite of many other considerations that could be adduced—are the rich societies and not the poor.

It is almost like a providential blessing that we, the rich countries, have found it in our heart at least to consider the Third World and to try to mitigate its poverty. In spite of the mixture of motives and the persistence of exploitative practices, I think that this fairly recent development in the outlook of the rich is an honorable one. And it could save us; for the poverty of the poor makes it in any case impossible for them successfully to adopt our technology. Of course, they often try to do so, and then have to bear the most dire consequences in terms of mass unemployment, mass migration into cities, rural decay and intolerable social tensions. They need, in fact, the very thing I am talking about, which we also need: a *different* kind of technology, a technology with a human face, which, instead of making human hands and brains redundant, helps them to become far more productive than they have ever been before.

As Gandhi said, the poor of the world cannot be helped by mass production, only by production by the masses. The system of mass production, based on sophisticated, highly capital-intensive, high energy-input dependent, and human labor-saving technology, presupposes that you are already rich, for a great deal of capital investment is needed to establish one single workplace. The system of *production by the masses* mobilizes the priceless resource which are possessed by all human beings, their clever brains and skillful hands, *and supports them with first-class tools*. The technology of *mass production* is inherently violent, ecologically damaging, self-defeating in terms of non-renewable resources, and stultifying for the human person. The technology of *production by the masses*, making use of the best of modern knowledge and experience, is conducive to decentralization, compatible with the laws of ecology, gentle in its use of scarce resources, and designed to serve the human person instead of making him the servant of machines. I have named it *intermediate technology* to signify that it is vastly superior to the primitive technology of bygone ages but at the same time much simpler, cheaper and freer than the super-technology of the rich. One can

also call it self-help technology, or democratic or people's technology—a technology to which everybody can gain admittance and which is not reserved to those already rich and powerful.

Although we are in possession of all requisite knowledge, it still requires a systematic, creative effort to bring this technology into active existence and make it generally visible and available. It is my experience that it is rather more difficult to recapture directness and simplicity than to advance in the direction of ever more sophistication and complexity. Any third-rate engineer or researcher can increase complexity; but it takes a certain flair of real insight to make things simple again. And this insight does not come easily to people who have allowed themselves to become alienated from real, productive work and from the self-balancing system of nature, which never fails to recognize measure and limitation. Any activity which fails to recognize a self-limiting principle is of the devil. In our work with the developing countries we are at least forced to recognize the limitations of poverty, and this work can therefore be a wholesome school for all of us in which, while genuinely trying to help others, we may also gain knowledge and experience of how to help ourselves. ∎

PART II

TOM BENDER

DETAIL FROM COVER DRAWING BY DIANE SCHATZ

A.T. DEFINITIONS

What is technology with a human face? What is appropriate *technology? Although no answer is final, and each answer varies with time, place and need, this collage of definitions gives some clues to its common features.*

You won't find a list of hardware or resources here. We've tried to avoid circular definitions which use tools to define the technology and vice versa, because not all solar collectors are appropriate and not all methane digesters are useful or even necessary. The important factor is always the context in which a tool or technique is used—scale, source, location and need. For more specific information about the range of different technologies (hardware and institutions alike) usually considered appropriate in various situations, see the Energy Primer *(40),* Soft Tech *(4),* The Appropriate Technology Sourcebook *(19), and* Rainbook *(49).*

The remainder of this section is about three essential qualities of appropriate technology: SIZE, SUSTAINABILITY *and* SIMPLICITY. *Appropriate technologies are scaled to suit the tasks being performed. They are themselves sustainable, or regenerating, and rely primarily on replenishable resources. And they use the simplest possible means to complete a task or achieve a goal.*

Common to all of these features is non-violence. *Appropriate technologies are above all gentle in their impact on their surroundings. They respect life rather than destroy it; they enliven rather than degrade. Appropriate technologies are also* decentralizing. *Their scale and lack of complexity tend to encourage localized approaches to problem solving. They enable regions, small towns, neighborhoods and individuals to reach towards a self-reliance (as opposed to self-sufficiency) that is rarely possible with centrally controlled machines, institutions or power sources.*□

The idea of appropriate technology is derived as much from a vision of a diversified society, of self-reliant and prosperous homes and communities, as it is from technology *per se*. Just building a few small-scale technology demonstration projects will not accomplish the broader goal of a more human technology. The idea of appropriate technology challenges the existing order of things, especially the values of centralized government, institutions and industries.

Wilson Clark
From "Big and/or Little? Search is on for the Right Technology," smithsonian magazine, 1976 (55)

The features of an appropriate technology are emerging in three areas: (1) a more integrated and steady-state relationship between the man-made and the natural environment so we do not overpower the self-healing capacities of natural systems to maintain themselves and support life; (2) developing social, economic and environmental diversity so that communities and regions can provide for many of their own needs without putting all their eggs in the single shrinking basket of imported and depleting resources; (3) creating and managing systems that require less capital, less outside energy, less machine watching and paper shuffling; but more personal involvement and direct production.

Sim Van der Ryn
From appropriate technology and state government, 1975 (59)

APPROPRIATE TECHNOLOGY demands that we ask, "Appropriate for what?" "Appropriate for whom?" No technology is value-free. Some technologies are appropriate for conditions of growth, vast resources and small populations. Others are appropriate for conditions of stability, plentiful labor and scarce resources. Some technologies increase the wealth of a few while impoverishing the many. Others tend to equalize wealth and power. Some technologies degrade and destroy the people using them, while others give opportunity for growth of skills, confidence and abilities. Some technologies produce goods, others produce good.

The existence of a technology does not require its use any more than the existence of a gun requires us to shoot. It only requires that we examine what happens if we don't use it and others do, and the effects its use would have on our lives.

Appropriate technology reminds us that before we choose our tools and techniques we must choose our dreams and values, for some technologies serve them, while others make them unobtainable.

Tom Bender
From RAINBOOK: RESOURCES IN APPROPRIATE TECHNOLOGY, 1977 (49)

A LIBERATED SOCIETY, I believe, will not want to negate technology precisely because it is liberated and can strike a balance. It may well want to assimilate the machine to artistic craftsmanship. By this I mean the machine will remove the toil from the productive process, leaving its artistic completion to man. The machine, in effect, will participate in human creativity. There is no reason why automatic, cybernated machinery cannot be used so that the finishing of products, especially those destined for personal use, is left to the community. The machine can absorb the toil involved in mining, smelting, transporting and shaping raw materials, leaving the final stages of artistry and craftsmanship to the individual. Most of the stones that make up a medieval cathedral were carefully squared and standardized to facilitate their laying and bonding—a thankless, repetitive and boring task that can now be done rapidly and effortlessly by modern machines. Once the stone blocks were set in place, the craftsmen made their appearance; toil was replaced by creative human work. In a liberated community the combination of industrial machines and the craftsman's tools could reach a degree of sophistication and of creative interdependence unparalleled in any period in human history. William Morris's vision of a return to craftsmanship would be freed of its nostalgic nuances. We could truly speak of a qualitatively new advance in technics—a technology for life.

Murray Bookchin
From TOWARDS A LIBERATORY TECHNOLOGY, 1971 (10)

NATURE OF TOOLS

By Ivan Illich
From TOOLS FOR CONVIVIALITY

A CONVIVIAL SOCIETY should be designed to allow all its members the most autonomous action by means of tools least controlled by others. People feel joy, as opposed to mere pleasure, to the extent that their activities are creative; while the growth of tools beyond a certain point increases regimentation, dependence, exploitation and impotence. I use the term "tool" broadly enough to include not only simple hardware such as drills, pots, syringes, brooms, building elements or motors, and not just large machines like cars or power stations; I also include among tools productive institutions such as factories that produce tangible commodities like corn flakes or electric current, and productive systems for intangible commodities such as those which produce "education," "health," "knowledge," or "decisions." I use this term because it allows me to subsume into one category all rationally designed devices, be they artifacts or rules, codes or operators, and to distinguish all these planned and engineered instrumentalities from other things such as basic food or implements, which in a given culture are not deemed to be subject to rationalization. School curricula or marriage laws are no less purposely shaped social devices than road networks.

Tools are intrinsic to social relationships. An individual relates himself in action to his society through the use of tools that he actively masters, or by which he is passively acted upon. To the degree that he masters his tools, he can invest the world with his meaning; to the degree that he is mastered by his tools, the shape of the tool determines his own self-image. Convivial tools are those which give each person who uses them the greatest opportunity to enrich the environment with the fruits of his or her vision. Industrial tools deny this possibility to those who use them, and they allow their designers to determine the meaning and expectations of others. Most tools today cannot be used in a convivial fashion.

Hand tools are those which adapt man's metabolic energy to a specific task. They can be multipurpose, like some primitive hammers or good modern pocket knives, or again they can be highly specific in design such as spindles, looms or pedal-driven sewing machines and dentists' drills. They can also be complex such as a transportation system built to get the most in mobility out of human energy—for instance, a bicycle system composed of a series of man-powered vehicles, such as pushcarts and three-wheel rickshas, with a corresponding road system equipped with repair stations and perhaps even covered roadways. Hand tools are mere transducers of the energy generated by man's extremities and fed by the intake of air and of nourishment.

Power tools are moved, at least partially, by energy converted outside the human body. Some of them act as amplifiers of human energy: the oxen pull the

"By reducing our expectations of machines we must guard against falling into the equally damaging rejection of all machines as if they were works of the devil," warns Ivan Illich in the sentence before this excerpt from Tools for Conviviality *(27), written in 1973. Here he defines "convivial" tools as those which enhance people's ability to be creative and relate to each other in friendship. Reprinted with permission from Harper and Row.*

plow, but man works with the oxen—the result is obtained by pooling the powers of beast and man. Power saws and motor pulleys are used in the same fashion. On the other hand, the energy used to steer a jet plane has ceased to be a significant fraction of its power output. The pilot is reduced to a mere operator guided by data which a computer digests for him. The machine needs him for lack of a better computer; or he is in the cockpit because the social control of unions over airplanes imposes his presence.

Tools foster conviviality to the extent to which they can be easily used, by anybody, as often or as seldom as desired, for the accomplishment of a purpose chosen by the user. The use of such tools by one person does not restrain another from using them equally. They do not require previous certification of the user. Their existence does not impose any obligation to use them. They allow the user to express his meaning in action.

Some institutions are structurally convivial tools. The telephone is an example. Anybody can dial the person of his choice if he can afford a coin. If untiring computers keep the lines occupied and thereby restrict the number of personal conversations, this is a misuse by the company of a license given so that persons can speak. The telephone lets anybody say what he wants to the person of his choice; he can conduct business, express love or pick a quarrel.

It is impossible for bureaucrats to define what people say to each other on the phone, even though they can interfere with—or protect—the privacy of their exchange.

Most hand tools lend themselves to convivial use unless they are artificially restricted through some institutional arrangements. They can be restricted by becoming the monopoly of one profession, as happens with dentist drills through the requirement of a license and with libraries or laboratories by placing them within schools. Also, tools can be purposely limited when simple pliers and screwdrivers are insufficient to repair modern cars. This institutional monopoly or manipulation usually constitutes an abuse and changes the nature of the tool as little as the nature of the knife is changed by its abuse for murder.

In principle the distinction between convivial and manipulatory tools is independent of the level of technology of the tool. What has been said of the telephone could be repeated point by point for the mails or for a typical Mexican market. Each is an institutional arrangement that maximizes liberty, even though in a broader context it can be abused for purposes of manipulation and control. The telephone is the result of advanced engineering; the mails require in principle little technology and considerable organization and scheduling; the Mexican market runs with minimum planning along customary patterns. ■

This is a new piece by Wendell Berry sent when we asked him for an article on appropriate technology and agriculture. Its premises can be applied to a.t. in general. It was written for Rodale Press for the first issue of their new magazine, The New Farm *(42). Reprinted with their permission.*

HORSE-DRAWN TOOLS AND THE DOCTRINE OF LABOR-SAVING

By Wendell Berry

Five years ago, when we enlarged our farm from about 12 acres to about 50, we saw that we had come to the limits of the equipment we had on hand: mainly a rotary tiller and a Gravely walking tractor; we had been borrowing a tractor and mower to clip our few acres of pasture. Now we would have perhaps 25 acres of pasture, 3 acres of hay, and the garden; and we would also be clearing some land and dragging the cut trees out for firewood. I thought for a while of buying a second-hand 8N Ford tractor, but decided finally to buy a team of horses instead.

I have several reasons for being glad that I did. One reason is that it started me thinking more particularly and carefully than before of the development of agricultural technology. I had learned to use a team when I was a boy, and then had learned to use the tractor equipment that replaced virtually all the horse and mule teams in this part of the country after World War II. Now I was turning around, as if in the middle of my own history, and taking up the old way again.

Buying and borrowing, I gathered up the equipment I needed to get started: wagon, manure spreader, mowing machine, disc, a one-row cultivating plow for the garden, assorted singletrees, doubletrees, breast yokes, etc. Most of these machines had been sitting idle for years. I put them

back into working shape and started using them. That was 1973. In the years since, I have bought a number of other horse-drawn tools, for myself and other people. My own outfit now includes a breaking plow, a two-horse riding cultivator, and a grain drill that I am in the process of rebuilding.

As I have repaired these old machines and used them, I have seen how well-designed and durable they are, and what good work they do. When the manufacturers modified them for use with tractors, they did not much improve either the machines or the quality of their work. (The tractor-mounted cultivator, in fact, is a much less flexible and sensitive tool than its horse-drawn forebear.) At the peak of their development, the old horse tools were excellent. The coming of the tractor made it possible for a farmer to do more work, but not better. And there comes a point, as we know, when *more* begins to imply *worse*. The mechanization of farming passed that point long ago—probably, or so I will argue, when it passed from horse-power to tractor-power. The increase of power has made it possible for one worker to crop an enormous acreage, but for this increased "efficiency" the country has paid a high price. From 1946 to 1976, because fewer people were needed, the farm population declined from 30 million to 9 million; the rapid movement of these millions into the cities greatly aggra-

GIGI COE

vated the complex of problems which we now call the "urban crisis," and the land is suffering for want of the care of those absent families. The coming of a tool, then, can be a cultural event of great influence and power. Once that is understood, it is no longer possible to be simple-minded about technological progress. It is no longer possible to ask, What is a good tool? without asking at the same time, How *well* does it work? and What is its influence?

One could say, as a sort of rule of thumb, that a good tool is one that makes it possible to work faster *and* better than before. At the point at which companies quit making them, the horse-drawn tools fulfilled both requirements. Consider, for example, the International High Gear No. 9 mowing machine. This is a horse-drawn mower that certainly improved on everything that came before it, from the scythe to previous machines in the International line. Up to that point, to cut fast and to cut well were two aspects of the same problem. Past that point the speed of the work could be increased, but not the quality.

I own one of these mowers. I have used it in my hayfield at the same time that a neighbor mowed there with a tractor mower; I have gone from my own freshly cut hayfield into others just mowed by tractors; and I can say unhesitatingly that, though the tractors do faster work, they do not do it better.

The same can be said, I think, of virtually all other tools: plows, cultivators, harrows, grain drills, seeders, spreaders, etc. Through the development of the standard two-horse equipment, quality and speed increased together; after that, the only increase has been in speed.

Moreover, as the speed has increased, care has tended to decline. For this, one's eyes can furnish ample evidence. But we have it also by the testimony of the equipment manufacturers themselves. Here, for example, is a quote from the public relations paper of one of the largest companies: "Today we have multi-row planters that slap in a crop in a hurry, putting down seed, fertilizer, insecticide and herbicide in one quick swipe across the field."

But good work and good workmanship cannot be accomplished by "slaps" and "swipes." Such language, and such behavior, seem to be derived from the he-man imagery of TV westerns, not from any known principles of good agriculture. What does the language of good agricultural workmanship sound like? Here is the voice of an old-time English farmworker and horseman, Harry Groom, as quoted in George Ewart Evans's *The Horse in the Furrow*: "It's all rush today. You hear a young chap say in the pub: 'I done 30 acres today.' But it ain't messed over, let alone done. You take the rolling, for instance. Two mile an hour is fast enough for a roll or a harrow. With a roll, the slower the better.

If you roll fast, the clods are not broken up, they're just pressed in further. Speed is everything [now]; just jump on the tractor and way across the field as if it's a dirt track. You see it when a farmer takes over a new farm: he goes in and plants straight-way, right out of the book. But if one of the old farmers took a new farm, and you walked round the land with him and asked him: 'What are you going to plant here and here?' he'd look at you some queer; because he wouldn't plant nothing much at first. He'd wait a bit and see what the land was like: he'd *prove* the land first. A good practical man would hold on for a few weeks, and get the feel of the land under his feet. He'd walk on it and feel it through his boots and see if it was in good heart, before he planted anything: he'd sow only when he knew what the land was fit for."

Granted that there is always plenty of room to disagree about farming methods, there is still no way to deny that in the first quote we have a description of careless farming, and in the second a description of a way of farming as careful—as knowing, skillful and loving—as any other kind of workmanship. The difference between the two is simply that the second considers where and how the machine is used, whereas the first considers only the machine. The first is the point of view of a man high up in the air-conditioned cab of a tractor described in an advertisement as "a beast that eats acres." The second is that of a man who has worked close to the ground in the open air of the field, who has studied the condition of the ground as he drove over it, and who has cared and thought about it.

If we had tools 35 years ago that made it possible to do farm work both faster and better than before, then why did we choose to go ahead and make them no longer better, but just bigger and bigger and faster and faster? It was, I think, because we were already allowing the wrong people to give the wrong answers to questions raised by the improved horse-drawn machines. Those machines, like the ones that followed them, were *labor savers*. They may seem old-timey in comparison to today's "acre-eaters," but when they came on the market they greatly increased the amount of work that one worker could do in a day. And so they confronted us with a critical question: How would we define labor-saving?

We defined it, or allowed it to be defined for us, as if it involved no human considerations at all, as if the labor to be "saved" were not human labor. We decided, in the language of some agricultural

experts, to look on technology as a "substitute for labor." Which means that we did not intend to "save" labor at all, but to *replace* it, and to *displace* the people who had once supplied it. We never asked what should be done with the "saved" labor; we let the "labor market" take care of that. Nor did we ask the larger questions of what values we should place upon people and their work and upon the land. It appears that we abandoned ourselves to a course of technological evolution, seemingly as unquestionable as natural law, which would value the development of machines far above the development of people.

And so it becomes clear that, by itself, my rule-of-thumb definition of a good tool is not enough. Even a tool that permits one to work both better and faster can cause bad results if its use is not directed by a benign and healthy social purpose. The coming of a tool, then, is not just a cultural event; it is also an historical crossroad—a point at which people must choose between two possibilities: to become more intensive or more extensive; to use the tool for quality or for quantity, for care or for speed.

In speaking of this as a choice, I am obviously assuming that the evolution of technology is *not* unquestionable or uncontrollable; that "progress" and the "labor market" do *not* represent anything so unyielding as natural law, but are aspects of an economy; and that any economy is in some sense a "managed" economy, managed by an intention to distribute the benefits of work, land and materials in a certain way. If those assumptions are correct, we are at liberty to undertake a kind of historical supposing, not meant, of course, to "change history" or "rewrite it," but to clarify somewhat this question of technological choice.

Suppose, then, that in 1945 we had valued the human life of farms and farm communities one percent more than we valued "economic growth" and technological progress. And suppose we had espoused the health of homes, farms, towns and cities with anything like the resolve and energy with which we built the "military-industrial complex." Suppose, in other words, that we had really meant what, all that time, most of us and most of our leaders were saying, and that we had really tried to live by the traditional values that we lip-served.

Then, it seems to me, we just might have accepted certain mechanical and economic limits. We might have used the improved horse-drawn tools, or even the small tractor equipment that followed, not to displace workers and decrease care and skill, but to

78

intensify production, increase care and skill, and widen the margins of leisure, pleasure and community life. We might, in other words, by limiting technology to a human or a democratic scale, have been able *to use the saved labor in the same places where we saved it*.

It is important to remember that "labor" is a very crude, industrial term, fitted to the huge economic structures, the dehumanized technology, and the abstract social organization of urban-industrial society. In such circumstances, "labor" means little more than the sum of two human quantities, human energy plus human time, which we identify as "man-hours." But the nearer home we put "labor" to work, and the smaller and more familiar we make its circumstances, the more we enlarge and complicate and enhance its meaning. At work in a factory, workers are only workers, units of human energy expending "man-hours" at a task set for them by strangers. At work in their own communities, on their own farms or in their own households or shops, workers are *never* only workers, but rather persons, relatives and neighbors. They work *for* those they work *among* or *with*. Moreover, it tends to be true that workers are independent in inverse proportion to the size of the circumstance in which they work. That is, the work of factory workers is ruled by the factory, whereas the work of housewives, small craftsmen or small farmers is ruled by their own morality and intelligence. And so, when workers work independently and at home, the society as a whole may lose something in the way of organizational efficiency and economies of scale. But it begins to *gain* values not so readily quantifiable in the fulfilled humanity of the workers, who then bring to their work, not just contracted quantities of "man-hours," but qualities such as independence, intelligence, judgment, pride, ambition, respect, love, reverence.

To put the matter in the most concrete terms, if the farm communities had been able to use the best horse-drawn tools to save labor in the true sense, then they might have used the saved time and energy, first of all, for leisure—something that technological progress has never given farmers. Second, they might have used it to improve their farms: to enrich the soil, prevent erosion, conserve water, put up better and more permanent fences and buildings; to practice forestry and its dependent crafts and economics; to plant orchards, vineyards, gardens of bush fruits; to plant market gardens; to improve pasture, breeding, husbandry and the de-

pendent enterprises of a local, small-herd livestock economy; to enlarge, diversify and deepen the economies of households and homesteads. Third, they might have used it to expand and improve the specialized crafts necessary to the health and beauty of communities: carpentry, masonry, leatherwork, cabinetwork, metalwork, pottery, etc. Fourth, they might have used it to improve the homelife and the home instruction of children, thereby preventing the tremendous hardships and expenses that the lack of same has now placed on schools, courts and jails.

It is probable also that, if we *had* followed such a course, we would have averted or greatly ameliorated the present shortages of energy and employment. The cities would not be overcrowded; the rates of crime and welfare dependency would be much lower; and the standards of industrial production would probably be higher.

I am aware that all this is exactly the sort of thinking that the technological evolutionists will dismiss as nostalgic or wishful. I mean it, however, not as a recommendation that we "return to the past," but as a criticism of the past; and my criticism is based on the assumption that we had in the past, and that we have now, a *choice* about how we should use technology and what we should use it for. As I understand it, this choice depends absolutely on our willingness to limit our desires as well as the scale and kind of technology we use to satisfy them. Without that willingness, there is no choice; we must simply abandon ourselves to whatever the technologists may discover to be possible.

The technological evolutionists, of course, do not accept that such a choice exists—undoubtedly because they resent the moral limits on their work that such a choice implies. They speak romantically of "man's destiny" to go on to bigger and bigger and more and more sophisticated machines. Or they take the opposite course and speak the tooth-and-claw language of Darwinism. "My principle," says ex-Secretary of Agriculture Earl Butz, "is *Adapt or die*."

I am, I think, as enthusiastic about the principle of adaptation as Mr. Butz. We differ only on the question of what should be adapted. He believes that we should adapt to the machines, that humans should be forced to conform to technological standards. I believe that the machines should be adapted to us—to serve our *human* needs as our history, our heritage, and our most generous hopes have defined them. ∎

80

SIZE

Over the last several decades, we have been told by planners, architects, economists, bureaucrats and big business that efficiency and productivity increase with greater size, and that this trend towards bigness was somehow beneficial and economical to everyone. Thus government, vertically integrated corporations, World Trade Centers and shopping malls all became common denominators in the size-equals-efficiency equation. Most often, the reason for greater size is cost. "Larger units keep the cost to the consumer down" is one standard justification, and the unwilling consumer becomes both the rationale and the victim of size. But nothing can erase the unease we feel alone in the cold, windy shadow of a skyscraper or the confusion we feel when our freedom, integrity and power are lost as we are absorbed into bigger and less responsive units.

"Small Is Beautiful" has become a catch phrase of people seeking more humanly scaled organizations, services, cities and machines. But, as E. F. Schumacher (52) and Leopold Kohr (30) note in the excerpts here, neither bigness nor smallness is the answer. Size needs to be appropriate to the task which is to be done. It is all relative. Crucial to finding the right balance is understanding when optimum size has been reached. Kohr's example here is the snail, who must balance the needs of a shell which can be lived in and carried at the same time.

The myth of the advantages of greater size is most prevalent in business and industry, where it generally means greater stability and concentration of power. Size rarely means greater efficiency or better service to the consumer, however. In his testimony before the Senate Select Committee on Small Business (57), Barry Stein notes that small firms are not only more efficient, they are also the most frequent source of technical innovation and are more responsive to social, economic and environmental changes. In short, they are more resilient.

The Briarpatch Network (12) in San Francisco is an example of the ways that sharing and cooperation can help small enterprises. This coalition of businesses and organizations can obtain many of the services they need (insurance, marketing and legal advice, bookkeeping, and capital) because of their association with each other. It is also an example of an appropriate middle way: Separately, each group would have difficulty purchasing the services that the larger network can afford to support. Together they have all that they need. □

Until recently, Leopold Kohr's work has received relatively little notice in this country. His essays, like those of Schumacher, appear regularly in Resurgence *(50). Kohr's most interesting work has been on optimal size in a new book titled* The Overdeveloped Nations: The Diseconomies of Scale *(30). The following piece is reprinted with permission from Schocken Books.*

ON SIZE

By Leopold Kohr
From THE OVERDEVELOPED NATIONS

IN A SUPERB STUDY on the interrelationship of growth and form, the great English biologist W. D'Arcy Thompson has shown why nature puts a stop to the growth of things once they have become large enough to fulfill their function. A tooth stops growing when it can effectively bite and chew. If it grew larger, it would violate its function. It would impede the organism it is meant to strengthen, and would have to be pulled out. Similarly a snail, after having added a number of widening rings to the delicate structure of its shell, suddenly brings its building activities, to which it has now become accustomed, to a stop. For, as D'Arcy Thompson points out, a single additional ring would increase the size of the shell sixteen times. Instead of adding to the welfare of the snail, it would burden it with such an excess of weight that any increase in its productivity would henceforth be absorbed by the task of coping with the added difficulties created by enlarging the shell beyond the limits set by its purpose. Moreover, since from that point on the problems of overgrowth begin to multiply at a geometric ratio while the snail's productive capacity can at best be extended at an arithmetic ratio, it follows that, once overgrowth sets in, the snail will never be able to catch up with the added problems created by it.

This is the fundamental philosophic reason why there is a limit to all growth. Though highly beneficial up to a certain point, beyond it, it not only becomes life's chief complexity; it becomes nature's principal tool by which it leads its organism to obsolescence and destruction.

But nature puts limits to growth not only on its biological organisms. In particular, it puts them also on its social organisms such as firms, cities or states.

The only difference is that natural instincts which guide the behavior of animals have become so dulled in the case of the human species as a result of educational sophistication and mechanical progress, that the social and economic implication of the limiting principle imposed on growth by function and form is not grasped with half the lucidity with which it has long been recognized in its biological context. Hence the constant emphasis with which the bulk of our political and economic theorists continue to stress the need for further growth in spite of the handwriting on the wall that has been warning the world since the 1940s that the problem for the developed part of the globe is no longer one of how to *foster* growth but how to *stop* it.

For just as the tooth or the shell of the snail are determined by the function they are meant to fulfill, so is the size of a firm, of a city or a state. The social function of a thing may have different aspects according to whether our philosophy is collectivist or individualist. However, since living standards are meaningful only in an individual context, we may dismiss the collectivist aspect from this analysis. And as to its individualistic implication, we need but re-state the previously cited Aristotelian idea that the function of the state is not expansion or glory but to provide the individual citizen with the essence of the good life, the *summum bonum*.

In other words, once a society has become large enough to furnish the convivial, economic, political and cultural needs of man in satisfactory, though not necessarily gluttonous, abundance—leisure to think, taverns to debate, churches to pray, universities to teach, theaters to inspire, the arts to enchant—further growth can no longer add to its basic purpose. We have reached the point of diminishing living standards. ∎

This testimony is based on a more extensive work that Barry Stein wrote in 1974 titled Size, Efficiency and Community Enterprise *(57), published by the Center for Community Economic Development (65). CCED has completed a number of studies on the economics of small enterprises and community development. The following piece is reprinted with permission from the author.*

SMALL BUSINESS

By Barry Stein
Testimony Before the Senate Select Committee
on Small Business, December 2, 1975

I AM PLEASED to comment in connection with the Small Business Committee's hearings on "The Role of Small Business in Our Society." I am especially pleased by the Committee's expressed interest in the relationship of small business to American communities, American ideals and values, and the quality of American life. I say that because ordinarily, the phrase "small business" is taken to refer merely to a quantitatively distinct subcategory of business in general. On this assumption, considerable attention is paid to size and scale and their relationship to economic efficiency and market opportunity. Policies are devised in compensation for the presumed disadvantages of small size, and government agencies and offices arise to direct and monitor the resulting programs. Useful though this may all be, it has less than the desired impact because it ultimately rests on a misconception. Small business differs not quantitatively but qualitatively from business in general.

Small business in reality refers to a fundamentally different sort of economic system—one comprising economic entities and relationships with distinctive qualities lacking in the giant firms at the core of our present economy. Although size of firm is one dimension differentiating these two economic orders, it is far from the only one. Indeed, it is in some ways the least fundamental. What, after all, is a large firm? In virtually every case, it consists of a whole cluster of separate facilities, whether plants, offices, laboratories, or distribution points, under unitary control. The classic firm of economic theory, a single plant or establishment owned and operated by some person or group, is in actuality the essence of what we now call "small business." The issues of size and control are therefore confounded, and it is essential to consider both in any discussion of *either* small or large business. The very same plant or retail store in a community, depending on whether it is owned by a local entrepreneur or an international conglomerate, shifts from classification as "small" business to "large." However, since that does not change the establishment, we may well ask what difference the distinction makes and to whom.

Grossly speaking, five categories of persons and groups are affected by every firm's activities: society at large, the local communities of operation, customers, employees, and controlling beneficiaries. Since each of these is in some way affected by changes in the size of the firm and its control structure, government must decide among the conflicting benefits and costs that alternative arrangements entail, and adjust the rules of the game to provide what is thought to be a fair and proper balance among them. I firmly believe that the present set of rules is seriously distorted by this standard on two accounts. First, there is a widespread (and incorrect) presumption that large size is necessary for productive efficiency, from which it follows that the special virtues of small business must be subsidized. Second, existing regulations essentially fail to give adequate consideration to the costs and benefits of alternative means for distributing control over firms, even though these have a marked impact on both social and individual welfare.

Taking size first, some facts may put the matter in perspective. I restrict my remarks to manufacturing, both because that is the sector in which economies of scale are most important and because my own work has focused on it. American manufacturing units are surprisingly small; mean employment per establishment in 1972 was only 43.8 persons. To some extent, that is the result of a large number of tiny businesses still classified as manufacturers. However, even if these figures are disaggregated, most plants are of modest size. American multi-unit firms comprise only about 3 percent of the total number of American companies, although they own 16 percent of all plants and contain 73 percent of total manufacturing employment. These are the giants of American industry, and here, if anywhere, we can expect to find large plants. Yet the separate manufacturing establishments owned and operated by these firms employed only 203 people on the average, and that figure, by the nature of the data, is probably exaggerated. Moreover, if plants in a very few industries such as automobiles, defense systems and large electrical machinery are excluded, the mean employment figure drops to about 100. I do not deny the existence of many very large plants. I do deny that they are typical; the size of *plants* is very different from the size of *firms*.

Of course, economies of scale do exist. However, my own extensive survey and the great majority of studies carried out by others, including agents of the Congress, show clearly that these economies are generally achieved in individual plants of modest size. The exceptions are the large assembly-line operations to which I have already referred. The more critical issue, therefore, concerns possible economies to be gained by operating more than one efficient size plant; that is, in multi-plant or con-

glomerate firms. Here also most studies are in basic agreement. In most cases multi-unit firms are not more economical than single-unit firms operating an equivalent facility. Furthermore, economies from scale of such functions as development or finance, or in specialized skills, can be gained by other means as well. For example, separate firms can offer such services; this is now widely true for computer facilities. Cooperatives or other shared activities can provide more than adequate scale for other purposes; small Swiss firms pool credit and financial resources. Or again, industry and trade associations are common devices to gain economies of promotion, training and standardization.

There are also diseconomies of scale. For example, both the generation of innovations and their application are more effective in small firms than in large. This effect is often particularly marked in concentrated industries, where major technical developments tend to come from smaller, peripheral firms. Similarly, small firms are more responsive to environmental, social and economic changes, both in accuracy of perception and response to it. This capacity, which is important to national purpose, is related to the most fundamental source of diseconomies; larger firms acquire rapidly increasing costs of coordination. As information needs to flow further and across more junctions, it becomes both less accurate and less timely. There is also a substantial human cost. Available data indicate clearly that as size increases, worker satisfaction and mental health decrease, and strikes, physical injury, and sabotage increase. Many large firms attempt to minimize these problems by decentralizing as much decision-making as possible. Although this is useful, it is not equivalent to fully independent units and in principle can only gain a fraction of the benefits.

In light of this it is natural to seek figures on the actual size of plant necessary to exhaust economies of scale in particular industries. Previous work in this direction has led to extremely inconsistent results. Recently a colleague (Mark Hodax) and I carried out a new series of calculations utilizing detailed comparisons of successive Censuses of Manufactures to determine the size of new plants entering particular markets characterized by four-digit Standard Industrial Classifications, which are moderately precise. Specifically, we determined the minimum size of plants entering 100 manufactured consumer goods markets between 1963 and 1967. In effect, our procedure estimates the smallest employment-size plants thought by business to be viable in the marketplace. Using other data, we have also estimated the capital asset requirements of such plants and the necessary consumer market or numbers of average consumers needed to justify them. The results are as follows:

1. Almost 70 percent of these industries involved entry plants with fewer than 250 employees; 44 percent required fewer than 100 employees.

2. If automobile-related products are excluded, about 71 percent of all consumer goods by value could be produced for a market of 1 million persons. Twenty percent could be produced even for a population of 200,000.

3. Correspondingly, about 70 percent of the 100 industries require a capital investment in plant of less than $1 million.

These data suggest an important conclusion. A large fraction of existing manufacturing capacity operated under the control of multi-unit firms could be converted to independent operation in the market without loss of economic effectiveness, and with possible gain. Although this is an issue essentially of control rather than size, I certainly do not suggest it as a strategy. It is not feasible, and the short-term results would be disastrous. However, there is every reason to encourage individual actions of that sort, since an economy with a smaller proportion of multi-plant firms would be far less concentrated than our present one; business and individual opportunity would increase; wealth, power and income would all be more equitably distributed; and government regulation of business could be greatly reduced. Policies for this purpose can be devised. I will offer some tentative suggestions later.

I wish also to call attention to the fact that the plant sizes resulting from these calculations are in general well within the category of small business as defined by the Small Business Administration. This applies not only to the minimum entry sizes computed for consumer goods but also to the mean size already existing in multi-plant firms. Although I have not carried out the detailed calculation, it is certain that a considerable share of present manufacturing already takes place in plants more consistent with small business than large, a fact masked by confounding size of firm, size of plant and issues of control. The very phrase "small business," therefore, creates in the mind an exceptionally misleading image. Such units are substantial and com-

plex, and contain adequate capacity to carry out most of the economic functions of American society.

So much for size and effectiveness. As to control, I have already noted that the data on multi-plant firms make it clear that in many or most cases, unitary control is not necessary to gain what small economies of multi-plant scale exist and that, indeed, it may generate diseconomies of its own. In short, efficiency is certainly not the main drive to growth, although American values and traditions make it mandatory that big business claim greater efficiency, because it has no other acceptable reason for existence. In consequence, large industry devotes much energy to rhetorical demonstrations of its economic effectiveness.

In reality, large firms exist for quite different but persuasive reasons that explain why their size tends to increase, and why so many persons, groups and institutions prefer to associate with them. First, large firms have power and visibility, in which employees, suppliers and customers all wish to share, since they are thought to confer status and other social benefits. For this reason, people will pay a premium, even explicitly, to gain them. Second, power provides security. Employees feel more secure than they would in smaller enterprises, whether efficient or not. Indeed, from an individual point of view, efficiency is irrelevant. What is relevant is income, position and continued assurance of both. This both requires and helps firms to accumulate power above that characteristic of a competitive market. This power in turn becomes the basis of large firms' impact both on the market and on society in general.

The tendency to grow also derives in part directly from a particular form of limited control. Incremental benefits from growth, particularly financial, can be captured in most firms by the few people (owners or managers) who exercise control over the distribution of those benefits. If those few people can gain a dollar of additional income or a step up the Fortune 500 ladder, on incremental sales of a million dollars, it is worth their while to do so, even though it represents an extremely inefficient use of resources. So long as the benefits need not be widely distributed, growth will still be sought.

These characteristics account for the fact that detailed studies of financial returns of corporations as a function of asset size show a steadily decreasing efficiency of asset use, as one would generally expect from the principle of diminishing returns. That is, the ratio of net benefits to assets employed falls as size increases. The critical issues once again are control and power, not efficiency.

The institution of small business, on the contrary, intrinsically distributes power and control more broadly; it helps increase distributional equity in the society. This applies both within firms and to firms as market units.

The institution of small business, on the contrary, intrinsically distributes power and control more broadly; it helps increase distributional equity in the society. This applies both within firms and to firms as market units.

Internally, small firms offer individuals a sense of the whole and a chance to see the value of their own efforts. They enable personal relationships with executives, owners and decision makers. In larger firms these things are generally not true—a situation which we know leads to feelings of apathy, alienation and lowered mental health. Smaller firms also provide both more visible and more genuinely accessible opportunities to individuals, and role models that are more realistic. For example, there is a greater proportion of women managers in small than large firms. Problems related to productivity and the quality of work life are increasingly apparent, particularly in large firms. These problems can be ameliorated by appropriately combining a broader distribution of influence within the firm, with direct rewards. However, these arrangements in principle are more consistent with smaller firms and easier to apply to them.

Externally, of course, a system of independent and modest-size firms regenerates a healthy market, consistent with American political ideals that depend critically on the absence of large centers of private power. There would also then exist a greater array of perceived opportunities for individuals or groups to launch a business of their own, an established American way of increasing one's status and mobility. A system composed of relatively small organizations is demonstrably more accessible and can readily be seen as within the grasp of any person or group with sufficient interest and capacity, unlike the very large entities which presently populate our industrial universe.

A system of manageable and independent firms also helps provide the needed integration between economy and society, enhancing people's sense of coherence in their personal lives. This is doubly true where businesses participate in communities as interdependent systems, each adding strength and stability to the other. Such businesses are also more likely to multiply economic resources within the community. The relationship between firm and community becomes influential in decisions regarding environmental impact, hiring, purchases from local suppliers and use of local labor. Similarly, the distribution of benefits and opportunities between the firm and the community is likely to be more balanced, and a greater sense of local pride is derived from the presence of truly local firms, as against branches of externally controlled enterprises. In the former case, the firm necessarily considers its community. In the latter, the community is held hostage, since its interests are peripheral to powerful and externally controlled firms. There needs to be deliberate exploration of a variety of other control alternatives and support for those showing promise. For example, community development corporations, as I have argued elsewhere, combine better-than-typical performance with important community, social and human resource benefits.

In light of these conclusions, I propose for the Committee's consideration some possible changes in both legal structures and administrative practice. The aim is not sudden change; that would be neither appropriate nor most effective. Rather, it is to shift the course of future developments to increase the proportion, over time, of more independent, community-sustaining, opportunity-enhancing and effective smaller businesses.

1. Explicitly recognize size as a continuous variable; firms are not merely small or large. Tax abatements, preferential contracts, set-asides and subsidies should be redesigned along graduated lines up to fitting levels.

2. Such benefits should also take into account issues of control and distribution of equity as, for instance, in Iran. Single-plant firms should have preferential access to certain favored treatments. Firms whose ownership is locally or community-based should receive similar preference.

3. Both these arrangements should be tailored to the specific industry, and perhaps location, involved. In principle, differentiate firms as much as is practical along these several dimensions to relate rewards closely to desired actions and outcomes.

86

4. All government support for multi-plant firms should require justification in each case; there should be no such general or categorical subsidy.

5. In antitrust proceedings resulting in divestiture orders, the units to be divested should be functionally complete, and able to operate as firms in the market. Preference for purchase should be given to independent groups; those proposing to distribute control broadly within communities, or among workers or relatively disfavored persons, should have access to financial assistance.

6. All forms of existing government aid to business, notably from the SBA, should be readily available to community development corporations, cooperatives and the like. At present, this is not the case.

7. A detailed study should be made of the limits that existing laws set on the use of potentially valuable structural and work alternatives, especially in smaller businesses. The Fair Labor Standards Act of 1938, for example, prevents full application of innovations involving more flexible use of time.

In conclusion, arrangements in which large firms control many small plants that could be effectively independent permit certain beneficiaries of the large firms to derive extra benefits at the expense of a much larger but less powerful fraction of the population. As I have shown elsewhere, facilities are routinely shut down, or moved from one community to another, not because they are ineffective but because different arrangements generate preferential benefits. In this regard, New York City and Cornell, Wisconsin, have had similar problems. Our present economic structure does not achieve essential and widely sought social goals. We must provide other means. Support for independent, smaller-sized, single-plant, community-enhancing businesses is an important aspect of the necessary new policies. ∎

TOM BENDER

The Briarpatch Network has been so successful that it is difficult to understand why similar groups haven't developed in other places. This article was originally written for Co-Evolution Quarterly, Winter 1975 (16). *Since then the Briarpatch has expanded and is now employing three staff people. The insurance plan mentioned here is now in effect (covering such esoteric practices as acupuncture treatments). For more information about the businesses involved or for news from the Briarpatch, read* Briarpatch Review *(12), the first eight issues of which have been newly compiled into* The Briarpatch Book *(11). This article is reprinted with permission from* Co-Evolution Quarterly.

THE BRIARPATCH NETWORK

By Andy Alpine

THE BRIARPATCH is a support network of *people* in small businesses in the Greater San Francisco Bay Area who share similar values and business practices. I emphasize people because of the relationships, the source of the energy, the support, the love, comes from the people in the Briarpatch as individuals, as well as representatives of specific businesses. The main value that we share in common is that we are in business to learn about ourselves, about the community in which we live, about how we deal with customers and employees. The members of the Briarpatch Network are doing what they like to do—realizing that working at something you aren't pleased with is no way to spend time, much less a life. We daily experience what the Buddhists call Right Livelihood; a realization of self and a honing of the individual's talents through his/her work.

When we started talking about what values we shared, we used to say we were *not* in business merely for the economic return—to make a lot of money, but to learn about ourselves, etc. We are not against money or profit—it's just not our major reason for being in business. A few months ago we realized that we were talking about ourselves backwards—so now we place why we *are* in business

first and why not, second. We also feel that greed is not a necessary law of the universe and that sharing of skills, resources, labor, materials and ourselves is a natural thing to do, as is supporting others who think likewise. Furthermore, our belief in a simple lifestyle permits Briars to approach our businesses, our employees, our customers and our lives with joy and honesty. If we are not in business for a large monetary return, then our prices don't have to be as high, our employees can be treated more humanly, and our customers do not have to be looked upon as dollar bills walking through the door.

At first we spoke of ourselves as a support network of people in "alternative" small businesses. In discussing the Briarpatch with the general business community, as well as friends, we often found ourselves spending more time defining and defending the word "alternative" than dealing with what we were doing or trying to do, so we dropped it. As it now stands, the Briarpatch Network in the San Francisco Bay Area consists of members with a wide spectrum of economic and political viewpoints who find enough in common to value the relationships and the group as a whole.

In terms of business practices, the Network has

been fostering ownership participation, humanistic management, cooperative decision-making and publicly visible accounting, i.e., any Briars' books are open to the public. If a customer wants to know the reason for a particular item costing $1.75, a Briarpatch business is open to telling him/her why without being challenged to do so. Several individually owned Briarpatch businesses are now considering having the enterprise not only managed but owned by the people who work there. Briars who have had previous experience in dealing with the financial, legal and emotional questions that arise in such a situation are meeting with these other Briarpatch businesses to facilitate the transition.

That's what we are: lawyers, accountants, bankers, management, marketing and financial consultants, together with jugglers, bakers, psychologists, carpenters, veterinarians and insurance experts (to name just a few) offering each other our specific knowledge of the business world and beyond. We want fellow Briars to be successful, to enjoy their businesses, and we are there to support them.

When what is offered is more than knowledge and experience, members often trade skills on a barter basis, considering exchanges of energy as important as money.

In July of 1974 Dick Raymond of Portola Institute (who thought of the concept and the name of Briarpatch) and Louie Durham, formerly of San Francisco's Glide Memorial Church, pulled together a weekend meeting of people they knew who were involved in "Briarpatch type" businesses. About 25 of us, representing about as many businesses and organizations, met at a camp in the Santa Cruz mountains of California to discuss what we had in common. The conference was of great value in that the participants realized that there were others who supported their ideas. At the end of the conference, the consensus was for continuing involvement. For me, one of the most significant points raised was that success or failure of our businesses was determined by the participants in the business themselves, rather than the outside society. If for financial reasons The American Quilt Shop has to cease operation after one year, the outside world might say "Too bad! John and Mary wasted a year." However, the exact opposite might be true in terms of the knowledge, good times, insight, etc. that they experienced during that time—a great success.

So we decided to continue our involvement. A few weeks after the conference, about ten of us met in San Francisco and discussed the possibility of hiring a coordinator—someone who would keep us all informed of the activities of the other Briars. We needed about twelve individuals kicking in $25 a month for six months to pay a coordinator a subsistence wage for a trial period. It took a few more weeks to get that together, and in the interim my time became a little freer and I decided to take on the job of Briarpatch coordinator.

The following months were very exciting for me. Coming from a background of law and international business, I actually knew very little about the workings of small businesses and those who ran them (I had attended the Briarpatch weekend as a resource person for marketing). My first month was spent meeting and being with the 25 individuals that attended the Briarpatch camp, finding out about their individual businesses or organizations, their needs and just getting to know them, and them me. There were lots of walks in the park, sitting in backyards, learning how to bake bread in a commercial oven, talking about automobile repair, co-ops, collectives.

After a while the Briars told friends and business associates about our "network" and there were more people to meet with, get to know, share experiences with. While this was going on, Mike Phillips (financial consultant and author of *The Seven Laws of Money*), who had been involved from the

start of the Briarpatch concept with Dick Raymond, offered to meet with Briars on Wednesdays at the Pier 40 offices of Portola Institute on the San Francisco waterfront. The arrangement was that Briars who had specific questions about their business or who just needed some educated feedback would come to Pier 40 and speak with Mike or whoever else in the network might be of assistance. Appointments were set a few days in advance so that other Briarpatch members could be notified and come to offer their specific advice. Our Pier 40 "consulting sessions" have grown to include five Wednesday consultants attending on a regular basis and occasional resident experts sitting in; Mike's area is finance, Steve's is law, Rich's is management and personnel, Bonnie's is tax and bookkeeping and mine is marketing. We've also found that people who come seeking advice often provide excellent counsel to the person who has the appointment after theirs.

As the Briarpatch has expanded, the time for walks in the park with Briars has diminished. A great deal of the personal contact, and development of relationships, is by telephone, and this is okay. However, in line with our desire to keep in touch with the needs of the membership, Mike and I have started to spend two mornings a week visiting Briarpatch businesses who might not come to a Pier 40 meeting unless they had a specific problem. Spending an hour in someone's store, watching and listening can sometimes show a lot more than a cash flow ledger. The need for this was brought home when Gary and Oz invited me by their used bookstore roommate referral service one morning. Actually they were a bit insulted; I had not stopped by their store in the month or so since they opened. They choose not to have a phone and Briarpatch relationships have in some cases been phone oriented. It's incredible how exciting a community of people sharing where they are at can be—negative feedback, such as Gary and Oz's, presented in a constructive loving way, is indispensible. They also wanted me to come by to volunteer their energies in putting together the Briarpatch newsletter which will, on a tri-monthly basis, state what each Briar's individual needs are so that all of the Briarpatch (instead of only the coordinator) will know what to be looking for and be aware of.

A good example of the Briarpatch Network actually working occurred a few months ago with Play Experience. Play Experience designs playgrounds and shows the client community how to actually build the playground. Since neighbors often hit each other with hammers, no insurance company wanted to handle the necessary insurance policy. Werner, our Briarpatch insurance broker, went to 12 companies before he finally "persuaded" (using his company's reputation) one to take the account. The premium was $1500—an offer the insurance company figured Play Experience would never accept. However, they had to in order to carry out their work. They had only $500 in the bank, it was Tuesday, they were to build on Saturday. The Network was there with two Briars giving them $700 and $300 apiece at no interest for two months. It wasn't all on heart energy alone, since Play Experience had a good accounts receivable situation, but it sure did work. When I thanked one of the lenders, she thanked me more intensely; she had really gotten off on having the opportunity to lend the money she had just received as a tax rebate, to re-energize it.

In addition to our social gatherings, we also present periodic workshops. Our first workshop was in October 1974, a few months after we had gotten started. It actually was a way for those of us who were starting to relate around Briarpatch concepts to come together. It was a day-long workshop on tax, legal, accounting and insurance questions facing a small business or organization. Several of our professionals made presentations and answered questions; we ate, listened to Mike lay out the *Seven Laws of Money*, did some Sufi dancing, played some New Games and welcomed the first rain of the season. Since then The Numbers Game presented another day-long workshop specifically on tax and bookkeeping, and in July, together with the New Dimensions Foundation, Briarpatch put together an evening program on marketing and advertising. Participation has been quite good, with 50 people attending the first two workshops and 100 attending the advertising seminar. Due to the interest in advertising we'll be presenting, later this fall, a series of four Tuesday evening seminars covering media, public relations, direct mail and graphics, followed by a day-long work session. Workshops on silk-screening, aggression and competition are also being planned. There's a lot of talent within the Network, and these workshops and seminars are a way of spreading the wealth.

One project that is coming to fruition this fall is a group health insurance plan for Briarpatch members. This plan will include holistic approaches as well as standard medical treatment, and we're confident that Werner, our insurance broker, will put together quite an innovative package for Briars. Another member, Kathleen, has offered to compile a list of sympathetic medical practitioners and to handle the referral needs and implementation of such a plan. Kathleen is at U.C. Med. Center and in touch with both sides of the doctor-patient relationship.

Since I just mentioned the concept of Briarpatch members, I should relate how someone becomes a member: We've created the concept of "Briarship pledge"—this is an amount of money that an individual or business can afford to give the Network for a six-month period of time. Work or material exchanges have also been considered a valid Briarship pledge, and we've received such energy as computerization of our mailing list, printing of our envelopes, 100 modular wooden boxes and reduced rate veterinarian care for the coordinator's (my) animals, to name just a few. The range of monetary pledges has been between $30 and $120. It is important to us that the Briarpatch Network be supported by the members themselves and that the need to go for outside funds be kept at a minimum. Our overhead is low; in my capacity as Briarpatch coordinator, I'm the only paid employee and our only other expenses are telephones, mailings, mimeographing and gas. I've been drawing $300 a month and about $50 of that goes to expenses. We've made it so far without too much difficulty, and as the Network membership expands, most of the financial insecurity should disappear. . . .

The Briarpatch Network has been fun, and it looks like we'll be playing together for a while. The Network offers Briars the opportunity to give time, labor, love, energy, materials, skills and money to other Briars, and of course, to receive. It works! ∎

WHAT I WISH to emphasize is the *duality* of the human requirement when it comes to the question of *size*: there is no single answer. For his different purposes man needs many different structures, both small ones and large ones, some exclusive and some comprehensive. . . . Today, we suffer from an almost universal idolatry of giantism. It is therefore necessary to insist on the virtues of smallness —where this applies. If there were a prevailing idolatry of smallness, irrespective of subject or purpose, one would have to try to exercise influence in the opposite direction.

The question of scale might be put in another way: what is needed in all these matters is to discriminate, to get things sorted out. For every activity there is a certain appropriate scale, and the more active and intimate the activity, the smaller the number of people that can take part, the greater is the number of such relationship arrangements that need to be established. Take teaching: one listens to all sorts of extraordinary debates about the superiority of a teaching machine over some other forms of teaching. Well, let us discriminate: what are we trying to teach? It then becomes immediately apparent that certain things can only be taught in a very intimate circle, whereas other things can obviously be taught *en masse*, via the air, via television, via teaching machines and so on. What scale is appropriate? It depends on what we are trying to do.

E.F. Schumacher
From SMALL IS BEAUTIFUL, 1973 (52)

ANCIL NANCE

92

SUSTAINABILITY

The lessons of ecological history show that if steep, erosion-prone slopes are plowed to the fences, and the same nutrient-robbing crops are planted year after year, something must give. Productivity declines and farmers fall into debt as the resource, which is the economic base, is destroyed. The same is true of forestry. Without conscientious reforestation and careful logging practices, both the resource and the community of loggers and millworkers eventually are undermined as trees are cut and the lumber companies move on.

Nothing is more fundamental than the interconnection between carefully sustained resources and the well-being of human communities. Yet with so much space to move in and a petroleum substitute for nearly everything, it's no surprise that this relationship began to erode. Now that space is tight and the substitute is running out, sustainability—the ability to sustain and be sustained—becomes more critical.

Understanding sustainability means learning to think in terms of interrelated systems. Every part of a system affects every other; when one part is weakened or damaged, the whole suffers. Sir Albert Howard, a British botanist and agriculturalist, based his work on this premise, researching the critical link between the health of the soil and the well-being of plants and animals. As Howard explains in the conclusion to Soil and Health (26), reprinted here, the maintenance of soil through frequent nutrient recycling is key to a permanently sustainable agriculture and to the stability of civilization.

In Forest Farming (20), Sholto Douglas and Robert Hart take this idea a step further, showing how trees, a renewable and sustainable resource, can be beneficially used as a food source and as a means to revitalize stripped and barren land.

Many people are finding ways to help sustain themselves without impairing natural self-healing cycles by recycling, growing their own food, and even composting their own sewage and generating their own electricity. The New Alchemy Institute (69) is one of the several research groups which has integrated these techniques in a wholistic approach to energy and food production. John Todd, one of the founders of the New Alchemists, describes here the philosophical base for their work. The term he uses for these approaches is "biotechnology." He is not saying—nor is any other similar group saying—that there is one right way to approach integrated systems. Indeed, his premise is that we need to get away from the homogeneity of our present culture into varied and diverse techniques of sustaining and nurturing our resources.

Energy technologies, as well, contribute to a more sustainable future. The sun, wind, water and biofuels can adequately provide for our primary energy needs, and form the basis for what Amory Lovins calls the soft path. The soft path, one which is inherently less exploitative and authoritarian, relies on energy conservation and renewable energy sources. □

Soil and Health (26) *was written in 1947 and remains a classic introduction to sustainable, organic ways of growing food and recycling "waste." Equally important, Howard relates the health of agriculture to the health of animals, people and nations. Reprinted here with permission from Devin-Adair Publishing Company.*

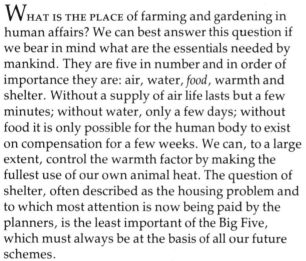

THE SOIL & HEALTH

By Sir Albert Howard
FROM THE SOIL & HEALTH

WHAT IS THE PLACE of farming and gardening in human affairs? We can best answer this question if we bear in mind what are the essentials needed by mankind. They are five in number and in order of importance they are: air, water, *food*, warmth and shelter. Without a supply of air life lasts but a few minutes; without water, only a few days; without food it is only possible for the human body to exist on compensation for a few weeks. We can, to a large extent, control the warmth factor by making the fullest use of our own animal heat. The question of shelter, often described as the housing problem and to which most attention is now being paid by the planners, is the least important of the Big Five, which must always be at the basis of all our future schemes.

Our food is produced for the most part by farmers and gardeners. It has been sadly neglected in the past, as will be clear to anyone who studies this book and its many implications. The essential things about food are three: (1) it must be grown in fertile soil, that is to say in soil well supplied with freshly prepared, high quality humus; (2) it must be fresh; (3) its cost must be stabilized in such a manner as to put an end to the constant fluctuations and steady rise in prices. All these things are possible once we increase the efficiency of the earth's green carpet—the machinery furnished by Nature for producing food. *The sun provides the energy for running this mechanism, so our power problem has been solved for us.* The sole food producing machine is the green leaf. This, again, is the gift of Providence. Mankind can increase the efficiency and output of this green carpet at least threefold by (1) the restoration and maintenance of the fertility of the soil on which it rests and (2) by providing varieties of crops which make the most of the sun's rays and the improved soil conditions. The former can be achieved by converting into humus the vast stores of vegetable and animal residues now largely running to waste: the latter by modern plant-breeding methods. Once we do this, all goes well. The roots are provided with a favorable climate and ample living space. The yield and quality of the produce go up by leaps and bounds: the danger of any shortage of food in the world disappears: the problem of price regulation is automatically solved.

We can check our food production methods by means of Nature's censors—the diseases of crops and livestock. Provided we prepare the soil for its manurial rights by suitable cultivation and subsoiling, and then faithfully comply with Nature's great

law of return by seeing to it that all available vegetable, animal and human wastes are converted
into humus in suitable heaps or pits outside the land
or in the soil itself by the processes of sheet-com-
posting, we shall soon find that many striking
things will begin to happen. The yield and quality
will rapidly improve: the crops will be able to resist
the onslaughts of parasites: *well-being and content-
ment, as well as the power to vanquish disease, will be
passed on to the livestock which consume them:* the
varieties of crops cultivated will not run out, but will
preserve their power of reproduction for a very long
time.

The objection of composting on the average farm
or market garden on the score of the dearness and
scarcity of labor is being removed by the mechanization of the manure heap. Several machines have already been devised which will assemble the compost heaps, turn them, and load the finished humus
on to suitable manure distributors. With the help of
one of these machines the cost per ton has already
been reduced to less than a quarter. This suggests
that mechanized organic farming and gardening is
certain to prove much cheaper than the methods
now in use, where the manurial rights of the soil
and of the crop are being largely evaded by substitutes in the shape of artificial manures. Large-scale
results coming in a growing torrent from all over the
world show that the ephemeral methods of manur-
ing, by means of chemicals and the resulting *survival
of the weakly plant bolstered up by poison sprays*, are
bound to be swept into the oblivion which they
merit.

The power to resist diseases, which organic
farming and gardening confer on the plant and on
the animal, is duly passed on to mankind. The evidence in favor of this view is rapidly growing. When
examples without end are available, showing how
most of the malnutrition, indisposition and actual
disease from which the population now suffers can
be replaced by robust health by merely living on the
fresh produce of fertile soil, it will be a simple matter
in any democratic country for the people to insist on
their birthright—fresh food from fertile soil—for
themselves and for their children. The various bodies which now stand in the way of progress will be
rapidly eliminated once their interests come in conflict with those of the electorate.

One of the great tasks before the world has been
outlined in this book. It is to *found our civilization on
a fresh basis—on the full utilization of the earth's green
carpet*. This will provide the food we need: it will
prevent much present-day disease at the source and
at the same time confer robust health and content-
ment on the population: it will do much to put an
end automatically to the remnants of *this age of ban-
ditry now coming to a disastrous close*. Does mankind
possess the understanding to grasp the possibilities
which this simple truth unfolds? If it does and if it
has the audacity and the courage to tread the new
road, then civilization will take a step forward and
the Solar Age will replace this era of rapacity which
is already entering into its twilight. ∎

AMONG THE MOST COMMON SIGHTS on our rides
from Yokohama to Tokyo, both within the city and
along the roads leading to the fields, starting early
in the morning, were the loads of night soil carried
on the shoulders of men and on the backs of
animals, but most commonly on strong carts drawn
by men, bearing six to ten tightly covered wooden
containers holding forty, sixty or more pounds each.
Strange as it may seem, there are not today and apparently never have been, even in the largest and
oldest cities of Japan, China or Korea, anything corresponding to the hydraulic systems of sewage disposal used now by western nations. Provision is
made for the removal of storm waters, but when I
asked my interpreter if it was not the custom of the
city during the winter months to discharge its night
soil into the sea, as a quicker and cheaper mode of
disposal, his reply came quick and sharp, "No, that
would be waste. We throw nothing away. It is
worth too much money." In such public places as
railway stations provision is made for saving, not
for wasting, and even along the country roads
screens invite the traveler to stop, primarily for
profit to the owner, more than for personal convenience.

F. H. King
From FARMERS OF FORTY CENTURIES, 1911 (29)

FOREST FARMING

By J. Sholto Douglas & Robert A. de J. Hart
From FOREST FARMING

THE MOST URGENT TASK facing mankind today is to find a comprehensive solution to the problems of hunger and malnutrition, with all the disease and misery that they involve, by methods that do not overburden stocks of non-renewable resources, such as oil and minerals used for fertilizers, and do not impoverish the environment.

Vast areas of the world which are at present unproductive or under-productive—savannahs and virgin grasslands, jungles and marshes, barren uplands and rough grazings, deserts and farmlands abandoned owing to erosion—could be brought to life and made more hospitable to human settlement. The know-how exists to make abundant contributions to man's food needs by methods combining scientific and technological research with traditional husbandry. The "tool" with the greatest potentials for feeding men and animals, for regenerating the soil, for restoring water systems, for controlling floods and droughts, for creating more benevolent micro-climates and more comfortable and stimulating living conditions for humanity, is the tree.

Of the world's surface, only eight to ten percent is at present used for food production. Pioneer agriculturists and scientists have demonstrated the feasibility of growing food-yielding trees in the most unlikely locations—rocky mountainsides and deserts with an annual rainfall of only two to four inches. With the aid of trees, at least three-quarters of the earth could supply human needs, not only of food but of clothing, fuel, shelter and other basic products. At the same time wildlife could be conserved, pollution decreased, and the beauty of many landscapes enhanced, with consequent moral, spiritual and cultural benefits.

The world's two greatest underdeveloped resources are the human capacities for creative fulfillment, which are thwarted by hunger, poverty, disease, violence and lack of educational opportunity, and the mineral-rich subsoils which can most efficiently be utilized by the powerful, questing roots of trees and other perennial plants.

The tree is a tool of almost unlimited versatility, the use of which does not, in general, involve technical skills beyond the capacity of the average human being. It can be grown in the form of extensive orchards or forests for the production of fruit, nuts and other edible and non-edible crops, or in the form of vast shelterbelts for the containment and reclamation of deserts. On the other hand, it can also be grown in small stands by the individual farmer or gardener who wishes to attain a measure of self-sufficiency.

The world food crisis is a problem which affects,

actually or potentially, every human being on earth.
Its solution depends, not only on governments and
international agencies, but on the efforts and initia-
tives of millions of private individuals.

The production of essential foods by conventional
methods of land use is lagging so far behind the
needs of the world's rapidly growing population
that even the advanced, industrialized, food-export-
ing countries are facing shortages of nutritional
factors that are vital for all-round positive health.
The toll of disease in the affluent countries which
can be attributed to diet deficiencies or toxic elements in food or the environment is becoming
comparable to the suffering caused by sheer malnutrition in the poorer countries. There are comparatively few people on earth whose health and
happiness could not be enhanced if they had access
to a comprehensive, balanced, natural diet consist-
ing largely of fresh products eaten direct from soil or
tree.

Imagination and boldness will be required to
bring into profitable bearing the huge neglected
and unexploited regions that now cover some three-
quarters of the land surface of the earth.

Apart from the fertile farmlands, the rest of the
world's inhabitable rural areas, considered from the
standpoint of their contributions to food and raw
material supplies, are used at the moment simply for
pastoral or low-density ranching activities, conventional forestry or orcharding, and various enterprises which contribute only marginally to the nourishment of the human race. In addition, some
of these activities are notoriously inefficient in land
use, output and operation.

The comprehensive answer to the problem of
these "delinquent landscapes," as a leading farmer
and forester graphically described them, is to incorporate them into integrated schemes of land use,
scientifically worked out to accord with soil and climatic factors. One of the most important factors in
such schemes should be massive tree plantings, for
trees can provide food and shelter for human
beings, livestock and crops and provide timber and
other products for building, fuel and industry; they
can heal erosion and control the movements of
water in the soil; they can purify polluted atmospheres and generally conserve the environment.

• • •

First and foremost, trees offer the possibility of
far higher food yields per acre. Whereas livestock
rearing in temperate regions produces an average of
about two hundredweight of meat per acre and
cereal growing an average of about one and a half

97

tons per acre, apple trees can yield at least seven tons per acre, while leguminous, bean-bearing trees, such as the honey locust, can provide fifteen to twenty tons of cereal-equivalent. In tropical areas, and under conditions of multiple cropping—where trees are interplanted with vines, vegetables or cereals—far higher yields can be expected.

• • •

Another outstanding advantage enjoyed by trees is that they can tolerate conditions in which every other form of food production would be impossible, such as steep, rocky mountainsides. Both olives and carobs, for example, can be planted in the clefts of rocks where no soil at all is apparent; their roots find the nutritional elements they require.

The ability of trees to tap deep underground water veins is a supreme asset in many of the world's arid areas. Certain trees have roots which can penetrate as much as several hundred feet into the subsoil and rocky sub-strata in their search for subterranean water. Drought-resistant trees such as the almond can survive and flourish in apparently waterless conditions where all other crops fail. With their capacity for storing water for long periods, some species of trees and shrubs can survive extended droughts that kill all other forms of vegetable life. Moreover the water drawn up by tree roots from the depths of the earth can also benefit their vegetable neighbors. Tree plantations are able to raise the entire water table over a wide area, thus bringing the possibilities of conventional agriculture and horticulture to regions where such activities had been considered out of the question.

• • •

Trees can be found which will tolerate both the rarefied air of great heights and the polluted atmosphere of industrial cities. In recent years, apple orchards have been established at heights of over 12,000 feet in Tibet, while J. Russell Smith, the American authority on tree crops, reported that, in the early years of this century, a honey locust had been seen bearing its long pods in foggy London. Better than any other crop, trees could supply the younger generation's demand for self-sufficiency. Many suburban areas could produce more food than open countryside stocked or cropped according to the conditions of orthodox agriculture if the full tree-growing potentialities of private gardens were exploited.

These facts suggest an answer to the world food crisis which can be applied to every part of the earth where trees will grow and animals exist; it is capable of operation on the smallest or the largest scale; it is far less demanding in energy, machinery and irrigation than conventional agriculture, and far from damaging the environment, it conserves and improves both soil and water resources and purifies the atmosphere.

This is the creation of balanced, ecological plant-and-animal communities, scientifically adapted to local climatic and soil conditions, and with species carefully selected for their favorable relationships with each other.

• • •

The general pattern of three-dimensional forestry is to have large belts or blocks of economic trees interspersed with narrower grazing strips of grasses or other herbage along which move herds of livestock, fed from the woodlands, and producing meat, milk, eggs, wool and other items. The system forms a natural biological cycle, into which man fits perfectly: he can eat the food harvested from the trees and the flesh or produce of the forest-fed livestock, or sell them. The manure of the animals is returned to the soil and encourages healthy and vigorous growth of plants, thus reducing the need for bought-in fertilizers to a minimum.

Three-dimensional forestry offers more than a system for satisfying man's basic needs of food, fuel and other essentials. It offers nothing less than a new way of life, which could provide rewarding and purposeful occupations for large populations. The drift of rural dwellers to the towns is fostering excessive urban expansion in many parts of the world, and leading to the mushrooming of shanty towns with their deplorable living conditions. By offering new schemes of land development the influx into the cities could be checked and new, vital rural civilizations and cultures created. People could return to the countryside to participate in agri-silvicultural activities which could provide profitable and meaningful occupations for thousands of workless individuals and families. Forest farming would provide many highly skilled jobs which could give the ambitious, technically minded young men and women of today status and satisfaction at least equivalent to any available to the industrial worker, and carried out in far more pleasant and healthy surroundings.

This is the conclusion of A Modest Proposal *written by John Todd of the New Alchemy Institute in 1970. It has been reprinted in several languages since that time and can be found complete in the* Book of the New Alchemists *(58)–a good place to go for more information about the "biotechnical" research of this group. Reprinted with their permission.*

BIOTECHNOLOGY

By John Todd

From A MODEST PROPOSAL

THE FIRST STEP toward countering homogeneity would be to create a biotechnology based upon an ecological ethic. This biotechnology would function at the lowest levels of society, providing inexpensive life-support bases for individual families, small farmers or communities who desire more independence and a way of life that restores rather than destroys this fragile planet. It would not be founded upon profit or efficiency considerations but on the philosophical view that all things are inter-connected and interdependent, and that the whole cannot be defined in monetary terms. Energy production, agriculture, landscapes and communities must be tied together within individual research programs, and each area should be considered as a unique entity worthy of study. From indigenous research projects would evolve a biotechnology that reflects the needs of each region and peoples. In this way it will be possible to have fantastically varied communities and landscapes, as each develops its own integration with the world around it.

The New Alchemists have begun studies to shape the skills needed to establish modern, relatively self-contained communities that capture their own power, grow their own foods and utilize their wastes. Smaller systems, which could be adopted by suburban dwellers or any peoples interested in significant personal changes on behalf of the environment, are in the early stages of development.

It is our feeling that the adoption of such support systems would increase individual independence and re-establish a much-needed link with the organic world. If such a bond were created, people may begin to relate more realistically with the larger world around them. It should not be overlooked that the findings of this new biotechnology, if widely adopted, would permit a nation to function more normally during periods of hardship.

The first task of this science is to explore the little-used, ancient, nonpolluting forms of energy such as the sun and the wind, and perhaps even the waves and water currents. On a small scale it will be possible to harness and create new potentials for "old-fashioned" forms of energy. Recent technological advances in energy conversion and storage ensure the likelihood of success. It will not be the task of this biotechnology to create energy systems capable of handling the needs of large cities or industries. However, its discoveries might make decentralization or a return to the countryside more appealing.

Research on new uses for the clean energy sources should be closely linked to an ecologically based agriculture and aquaculture. With this orientation there would be no imitation of the present agricultural sciences, which are basically committed to extending the large, unstable, monoculture agricultural enterprises and increasing specialization in food production. This bio-science should focus on

the creation of rich soils and the raising of high quality plants and animals together in sophisticated polyculture schemes. Its goal would be the founding of stable and beautiful agricultural environments. Rather than being geared toward displacing people from the land, it would attempt to reverse the trend by emphasizing intensive forms of culture, more varied diets on a local basis, and nurture a love of food producing and agronomy for its own sake. Noncommercial varieties of nutritious foods would be investigated, such as many of the nuts, weeds and berries, as each area of the country has its own set of neglected wild crops that would benefit from selective breeding.

Besides borrowing many of the research methods and insights of orthodox agriculture and ecology, food research programs should be tied directly with the community's needs and require little capital for adoption. Planting and tending of ecologically sophisticated agricultural plots will require far more training, knowledge and labor than is needed on contemporary farms. Yet this fact should prove an asset by raising the status of those involved. Person highly skilled in creating good soils and raising nourishing foods will be highly respected and emulated. To work toward restoring the landscapes

should become a major intellectual and physical pursuit of the present generation.

Those who teach, investigate and adopt the findings of this science will be deeply involved in most phases of raising of foods and the generation of power to meet their needs. Diverse multicrop food projects will add to community life by shaping exquisitely beautiful landscapes, which will be a great pleasure to live and work in. These microcosms may contain within themselves some of the seeds of change for the larger society around them.

It is proposed that throughout the country centers of research and education be established and modestly funded. They should not be controlled by the universities, although affiliation in some instances might add to the variety of inputs into their development. Apart from research funding, these centers should be relatively self-sufficient and the knowledge gained be made available to all through community extension on the part of the center's residents.

Our experiences in attempting to create a wholistic biotechnology for small communities has convinced us that many outstanding young scientists would gladly commit themselves to its development and success, for no less than our future may be at stake.

The full text of The Road Not Taken *(34) presents detailed and cogent arguments against the "hard path" (centralized fossil fuel and nuclear-based technologies) and for the "soft path" described here. In its entirety, it is a concise, sound case for these more benign energy technologies. Lovins' reasoning is documented in even greater detail in* Soft Energy Paths: Toward a Durable Peace *(35). His two earlier books,* World Energy Strategies *(36) and* Nonnuclear Futures *(37), with John Price, are useful sourcebooks for present energy options.* The Road Not Taken *was first published in* Foreign Affairs, *Fall, 1976 (23) and is reprinted here with the permission from the Council on Foreign Relations, Inc.*

ENERGY STRATEGY: THE ROAD NOT TAKEN?

By Amory Lovins

In 1976 some analysts still predicted economic calamity if the United States does not continue to consume twice the combined energy total for Africa, the rest of North and South America, and Asia except Japan. But what have more careful studies taught us about the scope for doing better with the energy we have? Since we can't keep the bathtub filled because the hot water keeps running out, do we really (as Malcolm MacEwen asks) need a bigger water heater, or could we do better with a cheap, low-technology plug?

IMPROVING ENERGY EFFICIENCY

There are two ways, divided by a somewhat fuzzy line, to do more with less energy. First, we can plug leaks and use thriftier technologies to produce exactly the same output of goods and services—and bads and nuisances—as before, substituting other resources (capital, design, management, care, etc.) for some of the energy we formerly used. When measures of this type use today's technologies, are advantageous today by conventional economic criteria, and have no significant effect on life-styles, they are called "technical fixes."

In addition, or instead, we can make and use a smaller quantity or a different mix of the outputs themselves, thus to some degree changing (or reflecting ulterior changes in) our life-styles. We might do this because of changes in personal values, rationing by price or otherwise, mandatory curtailments, or gentler inducements. Such "social changes" include car-pooling, smaller cars, mass transit, bicycles, walking, opening windows, dressing to suit the weather, and extensively recycling materials. Technical fixes, on the other hand, include thermal insulation, heat-pumps (devices like air conditioners which move heat around—often in either direction—rather than making it from scratch), more efficient furnaces and car engines, less overlighting and overventilation in commercial buildings, and recuperators for waste heat in industrial processes. Hundreds of technical and semi-technical analyses of both kinds of conservation have been done; in the last two years especially, much analytic progress has been made.

Theoretical analysis suggests that in the long term, technical fixes *alone* in the United States could probably improve energy efficiency by a factor of at least three or four[1]. A recent review of specific practical measures cogently argues that with only those technical fixes that could be implemented by about the turn of the century, we could nearly double the efficiency with which we use energy.[2] If that is correct, we could have steadily increasing economic activity with approximately constant primary energy use for the next few decades, thus stretching our present energy supplies rather than having to add massively to them. One careful comparison shows that *after* correcting for differences of climate, hydro-electric capacity, etc., Americans would still use about a third less energy than they do now if they were as efficient as the Swedes (who see much room for improvement in their own efficiency).[3] U.S. per capita energy intensity, too, is about twice that of West Germany in space heating, four times in transport.[4] Much of the difference is attributable to technical fixes.

Some technical fixes are already under way in the United States. Many factories have cut tens of percent off their fuel cost per unit output, often with practically no capital investment. New 1976 cars average 27 percent better mileage than 1974 models. And there is overwhelming evidence that technical fixes are generally much cheaper than increasing energy supply, quicker, safer, of more lasting benefit. They are also better for secure, broadly based employment using existing skills. Most energy conservation measures and the shifts of consumption which they occasion are relatively labor-intensive. Even making more energy-efficient home appliances is about twice as good for jobs as is building power stations: the latter is practically the least labor-intensive major investment in the whole economy.

The capital savings of conservation are particularly impressive. In the terms used above, the investments needed to *save* the equivalent of an extra barrel of oil per day are often zero to $3,500, generally under $8,000, and at most about $25,000—far less than the amounts needed to increase most kinds of energy supply. Indeed, to use energy efficiently in new buildings, especially commercial ones, the additional capital cost is often *negative*: savings on heating and cooling equipment more than pay for the other modifications.

To take one major area of potential saving, technical fixes in new buildings can save 50 percent or more in office buildings and 80 percent or more in some new houses.[5] A recent American Institute of Architects study concludes that, by 1990, improved design of new buildings and modification of old ones could save a third of our current *total* national energy use—and save money too. The payback time would be only half that of the alternative investment in increased energy supply, so the same capital could be used twice over. . .

So great is the scope for technical fixes now that we could spend several hundred billion dollars on them initially, plus several hundred million dollars per day—and still save money compared with increasing the supply! And we would still have the fuel (without the environmental and geopolitical problems of getting and using it). The barriers to far more efficient use of energy are not technical, nor in any fundamental sense economic. So why do we stand here confronted, as Pogo said, by insurmountable opportunities?

The answer—apart from poor information and ideological antipathy and rigidity—is a wide array of institutional barriers, including more than 3,000 conflicting and often obsolete building codes, an innovation-resistant building industry, lack of mechanisms to ease the transition from kinds of work that we no longer need to kinds we do need, opposition by strong unions to schemes that would transfer jobs from their members to larger numbers of less "skilled" workers, promotional utility rate structures, fee structures giving building engineers a fixed percentage of prices of heating and cooling equipment they install, inappropriate tax and mortgage policies, conflicting signals to consumers, misallocation of conservation's costs and benefits (builders vs. buyers, landlords vs. tenants, etc.), imperfect access to capital markets, fragmentation of government responsibility, etc.

Though economic answers are not always right answers, properly using the markets we have may be the greatest single step we could take toward a sustainable, humane energy future. The sound economic principles we need to apply include flat (even inverted) utility rate structures rather than discounts for large users, pricing energy according to what extra supplies will cost in the long run ("long-run marginal-cost pricing"), removing subsidies, assessing the total costs of energy-using purchases over their whole operating lifetimes ("life-cycle costing"), counting the costs of complete energy systems including all support and distribution systems, properly assessing and charging environmental costs, valuing assets by what it would cost to replace them, discounting appropriately, and encouraging competition through antitrust enforcement (including at least horizontal divestiture of giant energy corporations).

Such practicing of the market principles we

FIGURE I

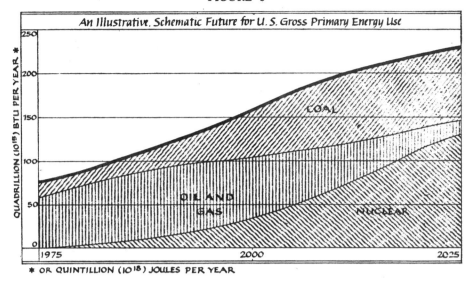

An Illustrative, Schematic Future for U. S. Gross Primary Energy Use

* OR QUINTILLION (10¹⁸) JOULES PER YEAR

preach could go very far to help us use energy efficiently and get it from sustainable sources. But just as clearly, there are things the market cannot do, like reforming building codes or utility practices. And whatever our means, there is room for differences of opinion about how far we can achieve the great theoretical potential for technical fixes. How far might we instead choose, or be driven to, some of the "social changes" mentioned earlier?

There is no definitive answer to this question—though it is arguable that if we are not clever enough to overcome the institutional barriers to implementing technical fixes, we shall certainly not be clever enough to overcome the more familiar but more formidable barriers to increasing energy supplies. My own view of the evidence is, first, that we are adaptable enough to use technical fixes *alone* to double, in the next few decades, the amount of social benefit we wring from each unit of end-use energy; and second, that value changes which could either replace or supplement those technical changes are also occurring rapidly. If either of these views is right, or if both are partly right, we should be able to double end-use efficiency by the turn of the century or shortly thereafter, with minor or no changes in life-styles or values save increasing comfort for modestly increasing numbers. Then over the period 2010-40, we should be able to shrink per capita primary energy use to perhaps a third or a quarter of today's.[6] (The former would put us at the per capita level of the wasteful, but hardly troglodytic, French.) Even in the case of four-fold shrinkage, the resulting society could be instantly recognizable to a visitor from the 1960s and need in no

sense to be a pastoralist's utopia—though that option would remain open to those who may desire it.

The long-term mix of technical fixes with structural and value changes in work, leisure, agriculture and industry will require much trial and error. It will take many years to make up our diverse minds about. It will not be easy—merely easier than not doing it. Meanwhile it is easy only to see what not to do.

If one assumes that by resolute technical fixes and modest social innovation we can double our end-use efficiency by shortly after 2000, then we could be twice as affluent as now with today's level of energy use, or as affluent as now while using only half the end-use energy we use today. Or we might be somewhere in between—significantly more affluent (and equitable) than today but with less end-use energy.

Many analysts now regard modest, zero or negative growth in our rate of energy use a realistic long-term goal. Present annual U.S. primary energy demand is about 75 quadrillion BTU ("quads"), and most official projections for 2000 envisage growth to 130-170 quads. However, recent work at the Institute for Energy Analysis, Oak Ridge, under the direction of Dr. Alvin Weinberg, suggests that standard projections of energy demand are far too high because they do not take account of changes in demographic and economic trends. In June 1976 the Institute considered that with a conservation program far more modest than that contemplated in this article, the likely range of U.S. primary energy demand in the year 2000 would be about 101-126

103

quads, with the lower end of the range more probable and end-use energy being about 60-65 quads. And, at the further end of the spectrum, projections for 2000 being considered by the "Demand Panel" of a major U.S. National Research Council study, as of mid-1976, ranged as low as about 54 quads of fuels (plus 16 of solar energy).

As the basis for a coherent alternative to the path shown in Figure 1, a primary energy demand of about 95 quads for 2000 is sketched in Figure 2. Total energy demand would gradually decline thereafter as inefficient buildings, machines, cars and energy systems are slowly modified or replaced Let us now explore the other ingredients of such a path—starting with the "soft" supply technologies which, spurned in Figure 1 as insignificant, now assume great importance.

SOFT ENERGY TECHNOLOGIES

There exists today a body of energy technologies that have certain specific features in common and that offer great technical, economic and political attractions, yet for which there is no generic term. For lack of a more satisfactory term, I shall call them "soft" technologies: a textural description, intended to mean not vague, mushy, speculative or ephemeral, but rather flexible, resilient, sustainable and benign. Energy paths dependent on soft technologies, illustrated in Figure 2, will be called "soft" energy paths, as the "hard" technologies constitute a "hard" path (in both senses). The distinction between hard and soft energy paths rests not on how

much energy is used, but on the technical and socio-political *structure* of the energy system, thus focusing our attention on consequent and crucial political differences.

In Figure 2, then, the social structure is significantly shaped by the rapid deployment of soft technologies. These are defined by five characteristics:
- They rely on renewable energy flows that are always there whether we use them or not, such as sun and wind and vegetation: on energy income, not on depletable energy capital.
- They are diverse, so that energy supply is an aggregate of very many individually modest contributions, each designed for maximum effectiveness in particular circumstances.
- They are flexible and relatively low-technology—which does not mean unsophisticated, but rather, easy to understand and use without esoteric skills, accessible rather than arcane.
- They are matched in *scale* and in geographic distribution to end-use needs, taking advantage of the free distribution of most natural energy flows.
- They are matched in *energy quality* to end-use needs: a key feature that deserves immediate explanation.

People do not want electricity or oil, nor such economic abstractions as "residential services," but rather comfortable rooms, light, vehicular motion, food, tables and other real things. Such end-use needs can be classified by the physical nature of the task to be done. In the United States today, about

FIGURE 2

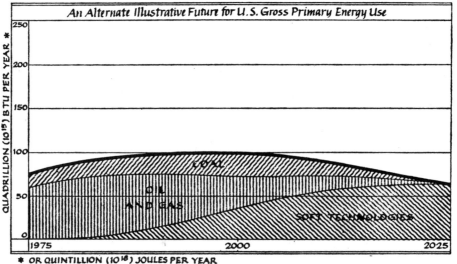

104

58 percent of all energy at the point of end use is required as heat, split roughly equally between temperatures above and below the boiling point of water. (In Western Europe the low-temperature heat alone is often a half of all end-use energy.) Another 38 percent of all U.S. end-use energy provides mechanical motion: 31 percent in vehicles, 3 percent in pipelines, 4 percent in industrial electric motors. The rest, a mere 4 percent of delivered energy, represents all lighting, electronics, telecommunications, electrometallurgy, electrochemistry, arc-welding, electric motors in home appliances and in railways, and similar end uses which now *require* electricity.

Some 8 percent of all our energy end use, then, requires electricity for purposes other than low-temperature heating and cooling. Yet, since we actually use electricity for many such low-grade purposes, it now meets 13 percent of our end-use needs—and its generation consumes 29 percent of our fossil fuels. A hard energy path would increase this 13 percent figure to 20-40 percent (depending on assumptions) by the year 2000, and far more thereafter. But this is wasteful because the laws of physics require, broadly speaking, that a power station change three units of fuel into two units of almost useless waste plus one unit of electricity. This electricity can do more difficult kinds of work than can the original fuel, but unless this extra quality and versatility are used to advantage, the costly process of upgrading the fuel—and losing two-thirds of it—is all for naught.

Plainly we are using premium fuels and electricity for many tasks for which their high energy quality is superfluous, wasteful and expensive, and a hard path would make this inelegant practice even more common. Where we want only to create temperature differences of tens of degrees, we should meet the need with sources whose potential is tens or hundreds of degrees, not with a flame temperature of thousands or a nuclear temperature of millions—like cutting butter with a chainsaw.

For some applications, electricity is appropriate and indispensable: electronics, smelting, subways, most lighting, some kinds of mechanical work, and a few more. But these uses are already oversupplied, and for the other, dominant uses remaining in our energy economy this special form of energy cannot give us our money's worth (in many parts of the United States today it already costs $50-120 per barrel-equivalent). Indeed, in probably no industrial country today can additional supplies of electricity

be used to thermodynamic advantage which would justify their high cost in money and fuels.

So limited are the U.S. end uses that really require electricity that by applying careful technical fixes to them we could reduce their 8 percent total to about 5 percent (mainly by reducing commercial overlighting), whereupon we could probably cover all those needs with present U.S. hydroelectric capacity plus the cogeneration capacity available in the mid-to-late 1980s.[7] Thus an affluent industrial economy could advantageously operate with no central power stations at all! In practice we would not necessarily want to go that far, at least not for a long time; but the possibility illustrates how far we are from supplying energy only in the quality needed for the task at hand.

A feature of soft technologies as essential as their fitting end-use needs (for a different reason) is their appropriate scale, which can achieve important types of economies not available to larger, more centralized systems. This is done in five ways, of which the first is reducing and sharing overheads. Roughly half your electricity bill is fixed distribution costs to pay the overheads of a sprawling energy system: transmission lines, transformers, cables, meters and people to read them, planners, headquarters, billing computers, interoffice memos, advertising agencies. For electrical and some fossil-fuel systems, distribution accounts for more than half of total capital cost, and administration for a significant fraction of total operating cost. Local or domestic energy systems can reduce or even eliminate these infrastructure costs. The resulting savings can far outweigh the extra costs of the dispersed maintenance infrastructure that the small systems require, particularly where that infrastructure already exists or can be shared (e.g., plumbers fixing solar heaters as well as sinks).

Small scale brings further savings by virtually eliminating distribution losses, which are cumulative and pervasive in centralized energy systems (particularly those using high-quality energy). Small systems also avoid direct diseconomies of scale, such as the frequent unreliability of large units and the related need to provide instant "spinning reserve" capacity on electrical grids to replace large stations that suddenly fail. Small systems with short lead times greatly reduce exposure to interest, escalation and mistimed demand forecasts—major indirect diseconomies of large scale.

The fifth type of economy available to small

systems arises from mass production. Consider, as Henrik Harboe suggests, the 100-odd million cars in this country. In round numbers, each car probably has an average cost of less than $4,000 and a shaft power over 100 kilowatts (134 horsepower). Presumably a good engineer could build a generator and upgrade an automobile engine to a reliable, 35-percent-efficient diesel at no greater total cost, yielding a mass-produced diesel generator unit costing less than $40 per kW. In contrast to the motive capacity in our central power stations—currently totaling about 1/40 as much as in our cars—costs perhaps ten times more per kW, partly because it is not mass-produced. It is not surprising that at least one foreign car maker hopes to go into the wind-machine and heat-pump business. Such a market can be entered incrementally, without the billions of dollars' investment required for, say, liquefying natural gas or gasifying coal. It may require a production philosophy oriented toward technical simplicity, low replacement cost, slow obsolescence, high reliability, high volume and low markup; but these are familiar concepts in mass production. Industrial resistance would presumably melt when—as with pollution-abatement equipment—the scope for profit was perceived.

This is not to say that all energy systems need be at domestic scale. For example, the medium scale of urban neighborhoods and rural villages offers fine prospects for solar collectors—especially for adding collectors to existing buildings of which some (perhaps with large flat roofs) can take excess collector area while others cannot take any. They could be joined via communal heat storage systems, saving on labor cost and on heat losses. The costly craftwork of remodeling existing systems—"backfitting" idiosyncratic houses with individual collectors—could thereby be greatly reduced. Despite these advantages, medium-scale solar technologies are currently receiving little attention apart from a condominium-village project on Vermont sponsored by the Department of Housing and Urban Development and the 100-dwelling-unit Mejannes-le-Clap project in France.

The schemes that dominate ERDA's solar research budget—such as making electricity from huge collectors in the desert, or from temperature differences in the oceans, or from Brooklyn Bridge-like satellites in outer space—do not satisfy our criteria, for they are ingenious high-technology ways to supply energy in a form and at a scale inappropriate to most end-use needs. Not all solar technologies are soft. Nor, for the same reason, is nuclear fusion a soft technology. But many genuine soft technologies are now available and are economic. What are some of them?

Solar heating and, imminently, cooling head the list. They are incrementally cheaper than electric heating, and far more inflation-proof, practically anywhere in the world. In the United States (with fairly high average sunlight levels), they are cheaper than present electric heating virtually anywhere, cheaper than oil heat in many parts, and cheaper than gas and coal in some. Even in the least favorable parts of the continental United States, far more sunlight falls on a typical building than is required to heat and cool it without supplement; whether this is considered economic depends on how the accounts are done. The difference in solar input between the most and least favorable parts of the lower 49 states is generally less than two-fold, and in cold regions, the long heating season can improve solar economics.

Ingenious ways of backfitting existing urban and rural buildings (even large commercial ones) or their neighborhoods with efficient and exceedingly reliable solar collectors are being rapidly developed in both the private and public sectors. In some recent projects, the lead time from ordering to operation has been only a few months. Good solar hardware, often modular, is going into pilot or full-scale production over the next few years, and will increasingly be integrated into buildings as a multipurpose structural element, thereby sharing costs. Such firms as Philips, Honeywell, Revere, Pittsburgh Plate Glass, and Owens-Illinois, plus many dozens of smaller firms, are applying their talents, with rapid and accelerating effect, to reducing unit costs and improving performance. Some novel types of very simple collectors with far lower costs also show promise in current experiments. Indeed, solar hardware per se is necessary only for backfitting existing buildings. If we build new buildings properly in the first place, they can use "passive" solar collectors—large south windows or glass-covered black south walls—rather than special collectors. If we did this to all new houses in the next 12 years, we would save about as much energy as we expect to recover from the Alaskan North Slope. [8]

Secondly, exciting developments in the conversion of agricultural, forestry and urban wastes to methanol and other liquid and gaseous fuels now offer practical, economically interesting technologies sufficient to run an efficient U.S. transport sector.[9] Some bacterial and enzymatic routes under study look even more promising, but presently proved processes already offer sizable contributions without the inevitable climatic constraints of fossil-fuel combustion. Organic conversion technologies must be sensitively integrated with agriculture and forestry so as not to deplete the soil; most current methods seem suitable in this respect, though they may change the farmer's priorities by making his whole yield of biomass (vegetable matter) saleable.

The required scale of organic conversion can be estimated. Each year the U.S. beer and wine industry, for example, microbiologically produces 5 percent as many gallons (not all alcohol, of course) as the U.S. oil industry produces gasoline. Gasoline has 1.5-2 times the fuel value of alcohol per gallon. Thus a conversion industry 10 to 14 times the scale (in gallons of fluid output per year) of our cellars and breweries would produce roughly one-third of the present gasoline requirements of the United States; if one assumes a transport sector with three times today's average efficiency—a reasonable estimate for early in the next century—then the whole of the transport needs could be met by organic conversion. The scale of effort required does not seem unreasonable, since it would replace in function half our refinery capacity.

Additional soft technologies include wind-hydraulic systems (especially those with a vertical axis), which already seem likely in many design studies to compete with nuclear power in much of North America and Western Europe. But wind is not restricted to making electricity: it can heat, pump, heat-pump or compress air. Solar process heat, too, is coming along rapidly as we learn to use the 5,800°C. potential of sunlight (much hotter than a boiler). Finally, high- and low-temperature solar collectors, organic converters, and wind machines can form symbiotic hybrid combinations more attractive than the separate components.

Energy storage is often said to be a major problem of energy-income technologies. But this "problem" is largely an artifact of trying to recentralize, upgrade and redistribute inherently diffuse energy flows. Directly storing sunlight or wind—or, for that matter, electricity from any source—is indeed difficult on a large scale. But it is easy if done on a scale and in an energy quality matched to most end-use needs. Daily, even seasonal, storage of low- and medium-temperature heat at the point of use is

straightforward with water tanks, rock beds, or perhaps fusible salts. Neighborhood heat storage is even cheaper. In industry, wind-generated compressed air can easily (and, with due care, safely) be stored to operate machinery: the technology is simple, cheap, reliable and highly developed. (Some cities even used to supply compressed air as a standard utility.) Installing pipes to distribute hot water (or compressed air) tends to be considerably cheaper than installing equivalent electric distribution capacity. Hydroelectricity is stored behind dams, and organic conversion yields readily stored liquid and gaseous fuels. On the whole, therefore, energy storage is much less of a problem in a soft energy economy than in a hard one.

Recent research suggests that a largely or wholly solar economy can be constructed in the United States with straightforward soft technologies that are now demonstrated and now economic or nearly economic.[10] Such a conceptual exercise does not require "exotic" methods such as sea-thermal, hot-dry-rock geothermal, cheap (perhaps organic) photovoltaic, or solar-thermal electric systems. If developed, as some probably will be, these technologies could be convenient, but they are in no way essential for an industrial society operating solely on energy income.

Figure 2 shows a plausible and realistic growth pattern, based on several detailed assessments, for soft technologies given aggressive support. The useful output from these technologies would overtake, starting in the 1990s, the output of nuclear electricity shown in even the most sanguine federal estimates. For illustration, Figure 2 shows soft technologies meeting virtually all energy needs in 2025, reflecting a judgment that a completely soft supply mix is practicable in the long run with or without the 2000-25 energy shrinkage shown. Though most technologists who have thought seriously about the matter will concede it conceptually, some may be uneasy about the details. Obviously the sketched curve is not definitive, for although the general direction of the soft path must be shaped soon, the details of the energy economy in 2025 would not be committed in this century. To a large extent, therefore, it is enough to ask yourself whether Figure 1 or 2 seems preferable in the 1975-2000 period.

A simple comparison may help. Roughly half, perhaps more, of the gross primary energy being produced in the hard path in 2025 is lost in conversions. A further appreciable fraction is lost in distribution. Delivered end-use energy is thus not vastly greater than in the soft path, where conversion and distribution losses have been all but eliminated. (What is lost can often be used locally for heating, and is renewable, not depletable.) But the soft path makes each unit of end-use energy perform several times as much social function as it would have done in the hard path; so in a conventional sense, social welfare in the soft path in 2025 is substantially greater than in the hard path at the same date.

• • •

SOME DEEPER ISSUES

Civilization in this country, according to some, would be inconceivable if we used only, say, half as much electricity as now. But that is what we did use in 1963, when we were at least half as civilized as now. What would life be like at the per capita levels of primary energy that we had in 1910 (about the present British level) but with doubled efficiency of energy use and with the important but not very energy-intensive amenities we lacked in 1910, such as telecommunications and modern medicine? Could it not be at least as agreeable as life today? Since the energy needed today to produce a unit of GNP varies more than 100-fold depending on what good or service is being produced, and since GNP in turn hardly measures social welfare, why must energy and welfare march forever in lockstep? Such questions today can be neither answered nor ignored.

Underlying energy choices are real but tacit choices of personal values. Those that make a high-energy society work are all too apparent. Those that could sustain life-styles of elegant frugality are not new; they are in the attic and could be dusted off and recycled. Such values as thrift, simplicity, diversity, neighborliness, humility and craftsmanship—perhaps most closely preserved in politically conservative communities—are already, as we see from the ballot box and the census, embodied in a substantial social movement, camouflaged by its very pervasiveness. Offered the choice freely and equitably, many people would choose, as Herman Daly puts it, "growth in things that really count rather than in things that are merely countable": choose not to transform, in Duane Elgin's phrase, "a rational concern for material well-being into an obsessive concern for unconscionable levels of material consumption."

Indeed, we are learning that many of the things we had taken to be the benefits of affluence are really remedial costs, incurred in the pursuit of benefits that might be obtainable in other ways without those costs. Thus much of our prized personal mobility is really involuntary traffic made necessary by the settlement patterns which cars create. Is that traffic a cost or a benefit?

Pricked by such doubts, our inflated craving for consumer ephemerals is giving way to a search for both personal and public purpose, to reexamination of the legitimacy of the industrial ethic. In the new age of scarcity, our ingenious strivings to substitute abstract (therefore limitless) wants for concrete (therefore reasonably bounded) needs no longer seem so virtuous. But where we used to accept unquestioningly the facile (and often self-serving) argument that traditional economic growth and distributional equity are inseparable, new moral and humane stirrings now are nudging us. We can now ask whether we are not already so wealthy that further growth, far from being essential to addressing our equity problems, is instead an excuse not to mobilize the compassion and commitment that could solve the same problems with or without the growth.

Finally, as national purpose and trust in institutions diminish, governments, striving to halt the drift, seek ever more outward control. We are becoming more uneasily aware of the nascent risk of what a Stanford Research Institute group has called ". . . 'friendly fascism'—a managed society which rules by a faceless and widely dispersed complex of warfare-welfare-industrial-communications-police bureaucracies with a technocratic ideology." In the sphere of politics as of personal values, could many strands of observable social change be converging on a profound cultural transformation whose implications we can only vaguely sense: one in which energy policy, as an integrating principle, could be catalytic?[11]

It is not my purpose here to resolve such questions—only to stress their relevance. Though fuzzy and unscientific, they are the beginning and end of any energy policy. Making values explicit is essential to preserving a society in which diversity of values can flourish.

Some people suppose that a soft energy path entails mainly social problems, a hard path mainly technical problems, so that since in the past we have been better at solving the technical problems, that is the kind we should prefer to incur now. But the hard path, too, involves difficult social problems.

We can no longer escape them; we must choose which kinds of social problems we want. The most important, difficult and neglected questions of energy strategy are not mainly technical or economic but rather social and ethical. They will pose a supreme challenge to the adaptability of democratic institutions and to the vitality of our spiritual life.

EXCLUSIVITY

These choices may seem abstract, but they are sharp, imminent and practical. We stand at a crossroads: without decisive action our options will slip away. Delay in energy conservation lets wasteful use run on so far that the logistical problems of catching up become insuperable. Delay in widely deploying diverse soft technologies pushes them so far into the future that there is no longer a credible fossil-fuel bridge to them: they must be well under way before the worst part of the oil-and-gas decline. Delay in building the fossil-fuel bridge makes it too tenuous: what the sophisticated coal technologies can give us, in particular, will no longer mesh with our pattern of transitional needs as oil and gas dwindle.

Yet these kinds of delay are exactly what we can expect if we continue to devote so much money, time, skill, fuel and political will to the hard technologies that are so demanding of them. Enterprises like nuclear power are not only unnecessary but a positive encumbrance for they prevent us, through logistical competition and cultural incompatibility, from pursuing the tasks of a soft path at a high enough priority to make them work together properly. A hard path can make the attainment of a soft path prohibitively difficult, both by starving its components into garbled and incoherent fragments and by changing social structures and values in a way that makes the innovations of a soft path more painful to envisage and to achieve. As a nation, therefore, we must choose one path before they diverge much further. Indeed, one of the infinite variations on a soft path seems inevitable, either smoothly by choice now or disruptively by necessity later; and I fear that if we do not soon make the choice, growing tensions between rich and poor countries may destroy the conditions that now make smooth attainment of a soft path possible.

These conditions will not be repeated. Some people think we can use oil and gas to bridge to a coal and fission economy, then use that later, if we wish, to bridge to similarly costly technologies in the hazy future. But what if the bridge we are now on is the last one? Our past major transitions in energy supply were smooth because we subsidized them with cheap fossil fuels. Now our new energy supplies are ten or a hundred times more capital-intensive and will stay that way. If our future capital is generated by economic activity fueled by synthetic gas at $25 a barrel-equivalent, nuclear electricity at $60-120 a barrel-equivalent, and the like, and if the energy sector itself requires much of that capital just to maintain itself, will capital still be as cheap and plentiful as it is now, or will we have fallen into a "capital trap"? Wherever we make our present transition to, once we arrive we may be stuck there for a long time. Thus if neither the soft nor the hard path were preferable on cost or other grounds, we would still be wise to use our remaining cheap fossil fuels—sparingly—to finance a transition as nearly as possible straight to our ultimate energy-income sources. We shall not have another chance to get there. ■

1 American Institute of Physics Conference Proceedings No. 25, *Efficient Use of Energy*, New York: AIP, 1975; summarized in *Physics Today*, August 1975.
2 M. Ross and R. H. Williams, "Assessing the Potential for Fuel Conservation," forthcoming in *Technology Review*; see also L. Schipper, *Annual Review of Energy* 1:455-518 (1976).
3 L. Schipper and A. J. Lichtenberg, "Efficient Energy Use and Well-Being: The Swedish Example," LBL-4430 and ERG-76-09, Lawrence Berkeley Laboratory, April 1976.
4 R. L. Goen and R. K. White, "Comparison of Energy Consumption Between West Germany and the United States," Stanford Research Institute, Menlo Park, Calif., June 1975.
5 A. D. Little, Inc., "An Impact Assessment of ASHRAE Standard 90-75," report to FEA, C-78309, December 1975; J. E. Snell *et al*. (National Bureau of Standards), "Energy Conservation in Office Buildings: Some United States Examples," International CIB Symposium on Energy Conservation in the Built Environment (Building Research Establishment, Garston, Watford, England), April 1976; Owens-Corning-Fiberglas, "The Arkansas Story," 1975.
6 A calculation for Canada supports this view: A. B. Lovins, *Conserver Society Notes* (Science Council of Canada, Ottawa), May/June 1976, pp. 3-16.
7 The scale of potential conservation in this area is given in Ross and Williams, *op cit*.; the scale of potential cogeneration capacity is from McCracken *et al*., *op. cit*.
8 R. W. Bliss, *Bulletin of the Atomic Scientists*, March 1976, pp. 32-40.
9 A. D. Poole and R. H. Williams, *Bulletin of the Atomic Scientists*, May 1976, pp. 48-58.
10 For examples, see the Canadian computations in A. B. Lovins, *Conserver Society Notes*, *op. cit*.; Bent Sørensen's Danish estimates in *Science* 189: 255-60 (1975).
11 W. W. Harman, *An Incomplete Guide to the Future*, Stanford Alumni Association, 1976.

110

SIMPLICITY

'Tis a gift to be simple, 'tis a gift to be free,
'Tis a gift to come down where we ought to be.
And when we are within a place just right
We will be in the valley of love and delight.
—OLD SHAKER HYMN

Simplicity, like size, is relative. The distinction that Schumacher makes about poverty in The Conscious Culture of Poverty *(p. 43) is equally true of simplicity: living a simple life does not mean living a deprived life. By the same token, a simply engineered tool generally works more effectively than a complicated gadget. Understanding the difference between essentials and extras in one's personal life, and knowing which of our trappings are actually traps, gives a freedom of choice which is often foreclosed in a cluttered and complex world. As Richard Gregg writes in* Voluntary Simplicity, *it is a matter of determining what really isn't needed any more, limiting one's desires, and sharing with others in order to gain a vastly richer life.*

Choosing a simpler life is, in itself, a political statement. This excerpt from The Long-Legged House *is far more than a discussion of the virtues of "going back to the land." In it Wendell Berry describes both the personal and political side of the choice he made to move to the country. It is a cogent plea for the need to find one's place and to live there wholly and simply, whether in the city or in the country.*

Simple techniques are often the most enduring. Cheap energy and cheap materials made possible a modern architecture that is a complex solution to the age-old problem of creating protected, comfortable space. The ability to mechanically heat and cool buildings allowed us to forget simpler techniques such as natural shading, solar orientation and the use of local building materials. Traditional builders did not have the means to alter the climate artificially. They built with and of the land, incorporating into their designs an awareness of the natural constraints and opportunities of a particular location.

In this excerpt from Architecture for the Poor, *Hassan Fathy describes the ancient technique of building with mud brick. While looking for a way to build affordable, livable public housing in Egypt, Fathy had the wisdom to explore the resources of the land he lived in to find practical cheap building materials. In the process he enabled the revival of an art and skill which had been dangerously close to extinction.*

Translating that simple sensibility to our own culture, Malcolm Wells, in An Ecological Architecture Is Possible, *explains some of the ways in which more sensitivity to place and climate can become part of our present building practices.* □

VOLUNTARY SIMPLICITY

By Richard Gregg

Voluntary simplicity of living has been advocated and practiced by the founders of most of the great religions—Jesus, Buddha, Lao Tse, Moses and Mohammed—also by many saints and wise men such as St. Francis and John Woolman, the Hindu rishis, the Hebrew prophets, the Moslem sufis; by many artists and scientists; and by such great modern leaders as Lenin and Gandhi. It has been followed also by members of military armies and monastic orders—organizations which have had great and prolonged influence on the world.

Clearly, then, there is or has been some vitally important element in this observance. But the vast quantities of things given to us by modern mass production and commerce, the developments of science and the complexities of existence in modern industrialized countries have raised widespread doubts as to the validity of this practice and principle. Our present "mental climate" is not favorable either to a clear understanding of the value of simplicity or to its practice.

We are not here considering asceticism in the sense of a suppression of instincts. What we mean by voluntary simplicity is not so austere and rigid. Simplicity is a relative matter, depending on climate, customs, culture, the character of the individual. For example, in India, except for those who are trying to imitate Westerners, everyone, wealthy as well as poor, sits on the floor, and there are no chairs. A large number of Americans, poor as well as rich, think they have to own a motor car, and many others consider a telephone exceedingly important. What is simplicity for an American would be far from simple to a Chinese peasant.

Voluntary simplicity involves both inner and outer condition. It means singleness of purpose, sincerity and honesty within, as well as avoidance of exterior clutter, of many possessions irrelevant to the chief purpose of life. It means an ordering and guiding of our energy and our desires, a partial restraint in some directions in order to secure greater abundance of life in other directions. It involves a deliberate organization of life for a purpose.

Of course, as different people have different purposes in life, what is relevant to the purpose of one person might not be relevant to the purpose of another. Yet it is easy to see that our individual lives and community life would be much changed if everyone organized and graded and simplified his purposes so that one purpose would easily dominate all the others, and if each person then reorganized his outer life in accordance with this new arrangement of purposes—discarding possessions and activities irrelevant to the main purpose. The degree of simplification is a matter for each individual to settle for himself.

• • •

To those who say that machinery and the apparatus of living are merely instruments and devices which are without moral nature in themselves, but which can be used for either good or evil, I would point out that we are all influenced by the tools and means which we use. Again and again in the lives of individuals and of nations we see that when certain means are used vigorously, thoroughly and for a long time, those means assume the character and influence of an end in themselves. We become obsessed by our tools. The strong quantitative elements in science, machinery and money, and in their products, tend to make the thinking and life of

This essay is excerpted from one which first appeared in the Indian journal, Visva-Bharati Quarterly, *in August 1936. We found it in the September 4 and 11, 1974 issues of* Manas *(38). Like so many of the pieces in this book, it has appeared in several other places in a variety of forms since that time. During his lifetime, Gregg worked with Gandhi in India, which explains some of the orientation of this piece.*

those who use them mechanistic and divided. The relationships which science, machinery and money create give us more energy outwardly but they live upon and take away from us our inner energy.

We think that our machinery and technology will save us time and give us more leisure, but really they make life more crowded and hurried. When I install in my house a telephone, I think it will save me all the time and energy of going to market every day, and much going about for making petty inquiries and minor errands to those with whom I have dealings. True, I do use it for those purposes, but I also immediately expand the circle of my frequent contacts, and that anticipated leisure time rapidly is filled by telephone calls to me or with engagements I make by the use of it. The motor car has the same effect upon our domestic life. We are all covering much bigger territory than formerly, but the expected access of leisure is conspicuous by its absence. Indeed, where the motor cars are very numerous, you can now, at many times during the day, walk faster than you can go in a taxi or bus.

The mechanized countries are not the countries noted for their leisure. Any traveller to the Orient can testify that the tempo of life there is far more leisurely than it is in the industrialized West. To a lesser degree, the place to find relative leisure in the United States is not in the highly mechanized cities, but in the country.

Moreover, we continually overlook the fact that our obsession with machinery spoils our inner poise and sense of values, without which the time spared from necessitous toil ceases to be leisure and becomes time without meaning, or with sinister meaning—time to be "killed" by movies, radio or watching baseball games, or unemployment with its degradation of morale and personality.

• • •

It is often said that possessions are important because they enable the possessors thereby to enrich and enhance their personalities and characters. The claim is that by means of ownership the powers of self-direction and self-control inherent in personality become real. Property, they say, gives stability, security, independence, a real place in the larger life of the community, a feeling of responsibility, all of which are elements of vigorous personality.

Nevertheless, the greatest characters, those who have influenced the largest numbers of people for the longest time, have been people with extremely few possessions. The reason for this is something that we usually fail to realize, namely that the essence of personality does not lie in its isolated individuality, its separateness from other people, its uniqueness, but in its basis of relationships with other personalities. It is a capacity for friendship, for fellowship, for intercourse, for entering imaginatively into the lives of others. At its height it is a capacity for and exercise of love. Friendship and love do not require ownership of property for either their ordinary or their finest expression. Creativeness does not depend on possession. Intangible relationships are more important to the individual and to society than property is. It is true that a certain kind of pleasure and satisfaction come from acquiring mastery over material things, but that sort of power and that sort of satisfaction are not so secure, so permanent, so deep, so characteristic of mental and moral maturity as are some others. The most permanent, most secure and most satisfying sort of possession of things other than the materials needed for bodily life, lies not in physical control and power of exclusion but in intellectual, emotional and spiritual understanding and appreciation. This is especially clear in regard to beauty.

• • •

Those by whom simplicity is dreaded because it spells lack of comfort may be reminded that some voluntary suffering or discomfort is an inherent and necessary part of all creation, so that to avoid all voluntary suffering means the end of creativeness.

There is one further value to simplicity. It may be regarded as a mode of psychological hygiene. Just as eating too much is harmful to the body, even though the quality of the food eaten is excellent, so it seems that there may be a limit to the number of things or the amount of property which a person may own and yet keep himself psychologically healthy. The possession of many things and of great wealth creates so many possible choices and decisions to be made every day that it becomes a nervous strain. One effect of this upon the will, and hence upon success in life, was deftly stated by Confucius:

"Here is a man whose desires are few. In some things he will not be able to maintain his resolution but they will be few.

"Here is a man whose desires are many. In some things he will be able to maintain his resolution but they will be few."

If a person lives among great possessions, they constitute an environment which influences him. His sensitiveness to certain important human relations is apt to become clogged and dulled, his imagination in regard to the subtle but important elements of personal relationship or in regard to lives in circumstances less fortunate than his own is apt to become less active and less keen. This is not always the result, but the exception is rare. When enlarged to inter-group relationships this tends to create social misunderstandings and friction.

The athlete, in order to win his contest, strips off the non-essentials of clothing, is careful of what he eats, simplifies his life in a number of ways. Great achievements of the mind, of the imagination, and of the will also require similar discriminations and disciplines.

Observance of simplicity is a recognition of the fact that everyone is greatly influenced by his surroundings and all their subtle implications. The power of environment modifies all living organisms. Therefore each person will be wise to select and create deliberately such an immediate environment of home things as will influence his character in the direction which he deems most important and such as will make it easier for him to live in the way that he believes wisest. Simplicity gives him a certain kind of freedom and clearness of vision . . .

In regard to aesthetics, simplicity should not connote ugliness. The most beautiful and restful room I ever entered was in a Japanese country inn, without any furniture or pictures or applied ornaments. Its beauty lay in its wonderful proportions and the soft colors of unpainted wood beams, paper walls and straw matting. There can be beauty in complexity but complexity is not the essence of beauty. Harmony of line, proportion and color are much more important. In a sense, simplicity is an important element in all great art, for it means the removal of all details that are irrelevant to a given purpose. It is one of the arts within the great art of life. And perhaps the mind can be guided best if its activities are always kept organically related to the most important purposes in life.

If simplicity of living is a valid principle, there is one important precaution and condition of its application. I can explain it best by something which Mahatma Gandhi said to me. We were talking about simple living and I said that it was easy for me to give up most things but that I had a greedy mind and wanted to keep my many books. He said, "Then don't give them up. As long as you derive inner help and comfort from anything, you should keep it. If you were to give it up in a mood of self-sacrifice or out of a stern sense of duty, you would continue to want it back, and that unsatisfied want would make trouble for you. Only give up a thing when you want some other condition so much that the thing no longer has any attraction for you, or when it seems to interfere with that which is more greatly desired." It is interesting to note that this advice agrees with modern Western psychology of wishes and suppressed desires. This also substantiates what we said near the beginning of our discussion, that the application of the principle of simplicity is for each person or each family to work out sincerely for themselves. ∎

The piece here is excerpted from an essay in which Berry explores the various ways of living in a manner which upholds and furthers one's principles. It was written during the Vietnam War on the occasion of the jailing of one of his students for conscientious objection. The full piece can be found in The Long-Legged House *(8), which is, sadly, out of print but well worth seeking out. Reprinted here with permission from the author.*

ON CITIZENSHIP AND CONSCIENCE

By Wendell Berry
From THE LONG-LEGGED HOUSE

A POSSIBILITY, equally necessary, and in the long run richer in promise, is to remove oneself as far as possible from complicity in the evils one is protesting, and to discover alternative possibilities.

To make public protests against an evil, and yet live in dependence on and in support of the way of life that is the source of the evil, is an obvious contradiction and a dangerous one. If one disagrees with the nomadism and violence of our society, then one is under an obligation to take up some permanent dwelling place and cultivate the possibility of peace and harmlessness in it. If one deplores the destructiveness and wastefulness of the economy, then one is under an obligation to live as far out on the margin of the economy as one is able: to be economically independent of exploitive industries, to learn to need less, to waste less, to make things last, to give up meaningless luxuries, to understand and resist the language of salesmen and public relations experts, to see through attractive packages, to refuse to purchase fashion or glamour or prestige. If one feels endangered by meaninglessness, then one is under an obligation to refuse meaningless pleasure and to resist meaningless work, and to give up the moral comfort and the excuses of the mentality of specialization.

One way to do this—the way I understand—is to reject the dependences and the artificial needs of urban life, and to go into the countryside and make a home there in the fullest and most permanent sense: that is, live on and use and preserve and learn from and enrich and enjoy the land. I realize that to modern ears this sounds anachronistic and self-indulgent, but I believe on the ground of my experience that it is highly relevant, and that it offers the possibility of a coherent and particularized meaningfulness that is beyond the reach of the ways of life of "average Americans." My own plans have come to involve an idea of subsistence agriculture—which does not mean that I advocate the privation and extreme hardship usually associated with such an idea. It means, simply, that along with my other occupations I intend to raise on my own land enough food for my family. Within the obvious limitations, I want my home to be a self-sufficient place.

But isn't this merely a quaint affectation? And isn't it a retreat from the "modern world" and its demands, a way of "dropping out"? I don't think so At the very least, it is a way of dropping *in* to a concern for the health of the earth, which institutional and urban people have had at second hand at best, and mostly have not had at all. But the idea has

116

other far-reaching implications, in terms of both private benefits and public meanings. It is perhaps enough to summarize them here by saying that when one undertakes to live fully on and from the land the prevailing values are inverted: one's home becomes an occupation, a center of interest, not just a place to stay when there is no other place to go; work becomes a pleasure; the most menial task is dignified by its relation to a plan and a desire; one is less dependent on artificial pleasures, less eager to participate in the sterile nervous excitement of movement for its own sake; the elemental realities of seasons and weather affect one directly and become a source of interest in themselves; the relation of one's life to the life of the world is no longer taken for granted or ignored, but becomes an immediate and complex concern. In other words, one begins to stay at home for the same reasons that most people now go away.

I am writing with the assumption that this is only one of several possibilities, and that I am obligated to elaborate this particular one because it is the one that I know about and the one that is attractive to me. Many people would not want to live in this way, and not wanting to seems the best reason not to. For many others it is simply not a possibility. But for those with suitable inclinations and the necessary abilities it is perhaps an obligation.

The presence of a sizable number of people living in this way would, I think, have a profound influence on the life of the country and the world. They would augment the declining number of independent small landowners. By moving out into marginal areas abandoned by commercial agriculture, they would restore neglected and impoverished lands, and at the same time reduce the crowdedness of the cities. They would not live in abject dependence on institutions and corporations, hence could function as a corrective to the subservient and dependent mentality developing among government people and in the mass life of the cities. Their ownership would help to keep the land from being bought up by corporations. Over a number of years, by trial and error, they might invent a way of life that would be modest in its material means and necessities and yet rich in pleasures and meanings, kind to the land, intricately joined both to the human community and to the natural world—a life directly opposite to that which our institutions and corporations envision for us, but one which is more essential to the hope of peace than any international treaty. ■

117

This excerpt from Architecture for the Poor *(22) is only a small part of the fascinating account of Fathy's lifelong work designing simple, appropriate shelter for Egyptians. It is a fine model for the adaptation of traditional techniques the world over to our needs today. Reprinted with permission from University of Chicago Press and the author.*

BUILDING WITH MUD BRICK

By Hassan Fathy

From ARCHITECTURE FOR THE POOR

SURELY IT WAS an odd situation that every peasant in Egypt with so much as an acre of land to his name had a house, while landowners with a hundred acres or more could not afford one. But the peasant built his house out of mud, or mud bricks, which he dug out of the ground and dried in the sun. And here, in every hovel and tumbledown hut in Egypt, was the answer to my problem. Here, for years, for centuries, the peasant had been wisely and quietly exploiting the obvious building material, while we, with our modern school-learned ideas, never dreamed of using such a ludicrous substance as mud for so serious a creation as a house. But why not? Certainly, the peasant's houses might be cramped, dark, dirty and inconvenient, but this was no fault of the mud brick. There was nothing that could not be put right by good design and a broom. Why not use this heaven-sent material for our country houses? And why not, indeed, make the peasants' own houses better? Why should there be any difference between a peasant's house and a landowner's? Build both of mud brick, design both well, and both could afford their owners beauty and comfort.

So I started to design country houses in mud brick. I produced a number of designs, and in 1937

even held an exhibition in Mansoura and later in Cairo, when I delivered a lecture about my conception of the country house. From this lecture came some chances to build. These houses, mostly for rich clients, were certainly an improvement on the old town type of country house, but largely because they were more beautiful. In spite of their economical mud brick walls, they were not so very much cheaper than houses built of more conventional materials, because the timber for the roofs was expensive.

Soon afterwards the war started, and all building stopped. Steel and timber supplies were completely cut off, and the army requisitioned such materials as were already in the country. Yet, still obsessed by my desire to build in the country, I looked about for ways of getting round the shortage. At least I still had mud bricks! And then it occurred to me that, if I had mud bricks and nothing else, I was no worse off than my forefathers. Egypt had not always imported steel from Belgium and timber from Rumania, yet Egypt had always built houses. But how had they built? Walls, yes. I could build walls, too, but I had nothing to roof them with. Couldn't mud bricks be used to cover my houses on top? What

118

about some sort of vault?

Normally, to roof a room with a vault, the mason will get a carpenter to make a strong wooden centering which has to be removed when the vault is made; this is a complete wooden vault, running the full length of the room, held up by wooden props, and on which the courses of the masonry vault will rest while being laid.

But besides being elaborate and requiring special skill to insure that the voussoirs are pointing toward the center of the curve, this method of construction is beyond the means of peasants. It is the kind of thing used in building a bridge.

Then I remembered that the ancients had built vaults without such centering, and I thought I would try to do the same. About this time I was asked to do some designs for the Royal Society of Agriculture, and I incorporated my new ideas in these houses. I explained my wishes to the masons, and they attempted to put up my vaults without using centering. The vaults promptly fell down.

Repeated attempts brought no more success. It was clear that, if the ancients had known how to build vaults without centering, the secret had died with them.

My elder brother happened at that time to be a

director working on the Aswan Dam. He heard of my failure, listened sympathetically, and then remarked that the Nubians were, in fact, building vaults that stood up during construction without using any support at all, to roof their houses and mosques. I was immensely excited; perhaps, after all, the ancients had not taken their secret to their provoking vaulted tombs with them. Perhaps the answer to all my problems, the technique that would at last let me use the mud brick for every part of a house, was awaiting me in Nubia.

One morning in February 1941 I got off the train in Aswan, in company with a number of students and teachers from the School of Fine Arts. The students were making a study tour of archaeological sites, and I had seized the chance of going with them to see what was to be seen in Nubia.

My first impression was of the extremely undistinguished architecture of Aswan itself. A small provincial town, looking like a seedy Cairo transplanted to the country; the same pretentious façades, the same gaudy shop fronts, the same poor-relation, apologetic, would-be metropolitan air. A depressing eyesore, spoiling the lovely and dramatic scenery of the Second Cataract. There was nothing in Aswan

119

The highest goodness, water-like,
Does good to everything and goes
Unmurmuring to places men despise;
But so, is close in nature to the Way.

—LAO TZE

for me; certainly no sign of the rumored techniques I had come in search of. I was so disappointed that I nearly decided to stay in my hotel.

However, I made the trip across the river, for my brother had told me that I must look at the villages of the district rather than Aswan itself. On entering the first village, Gharb Aswan, I knew that I had found what I had come for.

It was a new world for me, a whole village of spacious, lovely, clean and harmonious houses, each more beautiful than the next. There was nothing else like it in Egypt; a village from some dream country, perhaps from a Hoggar hidden in the heart of the Great Sahara—whose architecture had been preserved for centuries uncontaminated by foreign influences, from Atlantis itself it could have been. Not a trace of the miserly huddle of the usual Egyptian village, but house after house, tall, easy, roofed cleanly with a brick vault, each house decorated individually and exquisitely around the doorway with claustrawork—moldings and tracery in mud.

I realized that I was looking at the living survivor of traditional Egyptian architecture, at a way of building that was a natural growth in the landscape, as much a part of it as the dom-palm tree of the district. It was like a vision of architecture before the Fall: before money, industry, greed and snobbery had severed architecture from its true roots in nature. . .

I tried to find somebody who could tell me where the builders lived who had created this village. Here I was less lucky; all the men seemed to live away from the place, working in towns. There were only women and children there, and they were too shy to talk. The girls simply ran away giggling, and I could get no information at all.

Back in Aswan, with my appetite whetted but by no means satisfied, I continued my search for a mason who knew the secret of building these vaults. In the hotel I chanced to talk to the waiter about my quest, and he told me that there were indeed masons living in Aswan, and he would put me in touch with one. There was not much work for a profes-

sional builder of mud brick houses, it seemed, because every man in a village, whatever his usual job, was able to run up a vaulted house for himself, so these few masons were employed by the inhabitants of the provincial towns like Aswan, who had lost the knack of building in the traditional way. However, there were a very few masons who built the vaults, and the waiter said he would introduce me to *moallem* Boghdadi Ahmed Ali, the oldest of them.

The next day our party went to see the Fatimid Cemetery at Aswan. This is a group of elaborate shrines, dating from the tenth century, built entirely in mud brick, where vaults and domes are employed with splendid assurance and style.

There is also, close to Aswan, the Monastery of St. Simeon, a Coptic building of the same period. Here too mud brick domes and vaults are employed, but the simplicity and humility of the monastic ideal is revealed in the architecture, which thus proves able to accommodate equally well the contrasting inspirations of the Moslem and Christian religion. Among other things, I noticed with great surprise and interest that the refectory held a broad gallery, supported entirely upon an ingenious system of main and secondary vaults to avoid a heavy filling between the curved surface of the vault and the horizontal floor above it. This showed argument that mud brick buildings could go up to two stories and still be strong enough to survive for a thousand years. I was getting more and more confirmation of my suspicions that the traditional materials and methods of the Egyptian peasant were more than fit for use by modern architects, and that the solution to Egypt's housing problem lay in Egypt's history.

There still remained, though, to learn how the local vaulting was done. I had been promised a meeting with this master mason, but he did not show up. Not till the very last moment of our visit, when we were actually on the platform waiting for the train, did Boghdadi Ahmed Ali finally arrive, and there, with the train shrieking impatiently, amid a hissing of steam and the clanking of coaches, the shouts of guard and passengers and lookers-on,

120

we had just time to shake hands and exchange addresses before I was borne away to Luxor.

This architectural excursion was, for me, a hunt after mud brick vaults. After Aswan, we went to Luxor, where I was especially pleased to examine the granaries of the Ramesseum—long vaulted storehouses, built of mud brick 3,400 years old. It seemed to be a fairly durable substance.

From Luxor we went to Touna el Gebel, where I found more vaults, 2,000 years old, one supporting an excellent staircase.

It is curious that in one short tour I had seen standing proof of the prevalence of vaulting throughout Egyptian history, yet, from what we had been taught in the School of Architecture, I might never have suspected anyone before the Romans knew how to build an arch. Archaeologists confine their attention to broken pots and effaced inscriptions, their austere discipline being enlivened from time to time by the discovery of a hoard of gold. But for architecture they have neither eyes nor time. They can miss architectural statements placed right under their noses—there are books which state that the Ancient Egyptians could not build domes, and I have seen an Ancient Egyptian dome in the tomb of Seneb, right in the middle of the cemetery at Giza. There can be no question but that the technique of building vaults and domes—in mud brick, too—was perfectly familiar to Egyptians in the twelfth dynasty.

When I was back in Cairo, I wrote immediately to Aswan for masons. There was no time to lose, for the Royal Society of Agriculture's farm was still roofless after the collapse of our first attempt at vaulting. In a few days I had met Abdu Hamed and Abdul Rahim Abu en Nur—masons from Aswan—and the next day they were working on the farm. From the very first moment that I met them, they gave promise of a new era in building, for when they were asked how they would prefer to be paid, by the day or by the job, they were too simple to see any difference. Now the average workman much prefers to be paid by the day, for then he can take

frequent rests, refresh himself with coffee every half hour or so, and spin out the work so that it will continue to be a source of income to him for many weeks. It never occurred to these Aswani masons, though, that there could be two times for finishing a job, dependent on the method of payment, and they simply said they would roof a room for 120 piastres. When asked how long it would take, they said ''One and a half days.''...

And in fact when they started working it took them exactly one and a half days to roof one room. The terms agreed, the masons asked us to make them

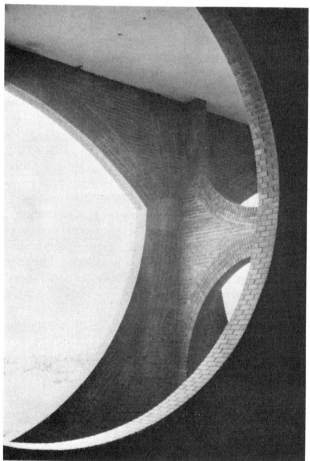

TOM BENDER

the special kind of bricks they used for vaults. These were made with more straw than usual, for lightness. They measured 25cm x 15cm x 5cm (10 in. x 6 in. x 2 in.) and were marked with two parallel diagonal grooves, drawn with the fingers from corner to corner of the largest face. These grooves were very important, for they enabled the bricks to stick to a muddy surface by suction. So we made the bricks and dried them, and a week later went down to the site. On our way down I noticed that the masons had no tools except for their adzes. I asked them, "Where are your trowels?" "We don't need trowels," they said. "The adze is enough."

At the scene of our failure the walls were still standing although our attempt at a vault had collapsed. In each room there were two side walls, three meters apart, and an end wall somewhat higher against which the vault was to be built. The masons laid a couple of planks across the side walls, close to the end wall, got up on them, took up handfuls of mud, and roughly outlined an arch by plastering the mud onto the end wall. They used no measure or instrument, but by eye alone traced a perfect parabola, with its ends upon the side walls. Then, with the adze, they trimmed the mud plaster to give it a sharper outline.

Next, one at each side, they began to lay the bricks. The first brick was stood on its end on the side wall, the grooved face flat against the mud plaster on the end wall, and hammered well into this plaster. Then the mason took some mud and against the foot of this brick made a little wedge-shaped packing, so that the next course would lean slightly towards the end wall instead of standing up straight. In order to break the line of the joints between the bricks the second course started with a half-brick, on the top end of which stood a whole brick. If the joints are in a straight line, the strength of the vault is reduced and it may collapse. The mason now put in more mud packing against this second course, so that the third course would incline even more acutely from the vertical. In this way the two masons gradually built the inclined courses out, each one rising a little higher round the outline of the arch, till the two curved lines of brick met at the top. As they built each complete course, the masons were careful to insert in the gaps between the bricks composing the course (in the extrados of the voussoirs) dry packing such as stones or broken pottery. It is most important that no mud mortar be put between the ends of the bricks in each course, for mud can shrink by up to 37 percent in volume,

and such shrinkage will seriously distort the parabola, so that the vault may collapse. The ends of the bricks must touch one another dry, with no mortar. At this stage the nascent vault was six brick-thicknesses long at the bottom and only one brick-thickness long at the top, so that it appeared to be leaning at a considerable angle against the end wall. Thus it presented an inclined face to lay the succeeding courses upon, so that the bricks would have plenty of support; this inclination, even without the two grooves, stopped the brick from dropping off, as might a smooth brick on a vertical face.

Thus the whole vault could be built straight out in the air, with no support or centering, with no instrument, with no drawn plan; there were just two masons standing on a plank and a boy underneath tossing up the bricks, which the masons caught dexterously in the air, then casually placed on the mud and tapped home with their adzes. It was so unbelievably simple. They worked rapidly and unconcernedly, with never a thought that what they were doing was quite a remarkable work of engineering, for these masons were working according to the laws of statics and the science of the resistance of materials with extraordinary intuitive understanding. Earth bricks cannot take bending and sheering; so the vault is made in the shape of a parabola conforming with the shape of the bending moment diagrams, thus eliminating all bending and allowing the material to work only under compression. In this way it became possible to construct the roof with the same earth bricks as for the walls. Indeed, to span three meters in mud brick is as great a technical feat, and produces the same sense of achievement, as spanning thirty meters in concrete.

The simplicity and naturalness of the method quite entranced me. Engineers and architects concerned with cheap ways of building for the masses had devised all sorts of complicated methods for constructing vaults and domes. Their problem was to keep the components in place until the structure was completed, and their solutions had ranged from odd-shaped bricks like bits of three-dimensional jigsaw puzzles, through every variety of scaffolding, to the extreme expedient of blowing up a large balloon in the shape of the required dome and spraying concrete onto that. But my builders needed nothing but an adze and a pair of hands.

Within a few days all the houses were roofed. Rooms, corridors, loggias were all covered with vaults and domes; the masons had solved every problem that had exercised me (even to building stairs). It only remained to go out and apply their methods throughout Egypt.

• • •

And this is the second great point about mud brick housing with vaulted roofs. Besides being cheap, it is also beautiful. It cannot help being beautiful, for the structure dictates the shapes and the material imposes the scale, every line respects the distribution of stresses, and the building takes on a satisfying and natural shape. Within the limits imposed by the resistance of materials—mud—and by the laws of statics, the architect finds himself suddenly free to shape space with his building, to enclose a volume of chaotic air and to bring it down to order and meaning to the scale of man, so that in his house at last there is no need of decoration put on afterward. The structural elements themselves provide unending interest for the eye. The vault, the dome, pendentive, squinches, arches and walls give the architect unlimited scope for a justified interplay of curved lines running in all directions with a harmonious passage from one to the other. ■

SOME WELL-KNOWN EXPERIMENTS show that it is quite possible to accumulate the solar heat by a simple apparatus, and thus to obtain a temperature which might be economically important even in the climate of Switzerland. Saussure (probably Horace Benedick de Saussure—1740-1799), by receiving the sun's rays in a nest of boxes blackened within and covered with glass, raised a thermometer enclosed in the inner box to the boiling point; and under the more powerful sun of the Cape of Good Hope, Sir John Herschel cooked the materials for a family dinner by a similar process, using, however, but a single box, surrounded with dry sand and covered with two glasses. Why should not so easy a method of economizing fuel be resorted to in Italy, and even in more northerly climates?

George P. Marsh
From MAN AND NATURE, 1894 (39)

We found this piece by Malcolm Wells in an old copy of the British magazine, Architectural Design *(2), but it was actually first printed in Rodale's* Environment Action Bulletin *(21) in January 1972. You can get a sense of how far things have come in six years from his comments about the dreams of solar energy. As he wrote recently, "At that time I was only 46 years old, so I had probably never seen a tree, a vegetable, a clothesline or a black car under an August sun." Malcolm now designs ecologically sound buildings himself, particularly underground ones (60). Reprinted with permission from the author.*

AN ECOLOGICALLY SOUND ARCHITECTURE IS POSSIBLE

By Malcolm B. Wells

Is it better to buy something already built in order to use up what we have (but also perpetuating an environmentally unsound building in operation), or is it better to build?

IF YOU CAN AFFORD to go off into the country somewhere and build an earthy-looking house in a beautiful, wild setting, the answer is easy. Just make sure your house is heated by nothing but sunlight and by the burning of wind-pruned tree limbs, see that all your wastes are allowed to rot into the land, and allow all rainwater from your roof to be absorbed by the deep humus around the building.

But the man who pays the greatest price for all my ecological mistakes is the guy I'd be leaving behind: the blue-collar worker, the black man, the Indian, or the Chicano. He couldn't build that woodland hideaway with its lush organic garden. He'd have to stay behind and breathe the worst of the air, drink the worst of the water, eat the worst of the food and live in the meanest of houses because I, who could perhaps afford to escape, refused to offer him the job or the friendship he needed to leave the city. And besides, there just isn't that much room left in paradise any more; not for 200 million of us there isn't. So we've got to solve this thing right here, wherever we are today, and we've got to realize that things like brotherhood and jobs are going to be as much a part of the solution as are waste-management and rebuilt cities.

SORTING OUT THE FACTS

It's tempting to start by talking about specifics, about not building in swamps or on flood plains, about things like the new "miracle" insulations, or percolation beds, or even earth-covered roofs, but unless we can reach some sort of agreement on the principles behind such architecture, we may in the name of ecology do more harm than good. Before we plunge we've got to wade, somehow, through all the baloney and sort out the facts.

Whether or not we're going to make all life extinct upon this planet within 30 or 40 years is, at this point at least, a matter of opinion. That we're already headed squarely in that direction is a matter of fact. All we need do to confirm it is look out the window, turn on the television set, or glance at a newspaper.

Another fact is that we have our priorities all wrong. We know, when we stop to think about it, that our basic needs are simple air, water, food and shelter. Period. We know that green plants are supposed to cover almost every square inch of the Earth. And we know that machines should be tucked away underground where they'll do the least amount of harm to the living land. But you'd never know we were aware of such facts when you see the way we build—even those of us who supposedly know better—or the way we do almost everything else. When you see what most of us are buying in the name of food, or where we're content to discharge our sewage, you'd think the basic priorities of life were coronary disease, artificiality and filth. If they are our priorities, then it hardly makes sense to be sitting here discussing the higher laws of architecture.

BUILDINGS DESTROY

Architecture, as it is practiced today, is a very messy business; a fact not as readily apparent to the casual observer of architects' working clothes as it is to, say, the casual observer of butchers' aprons or surgeons' gowns. But from ground-breaking day until the final demolition of a building—usually years after it has lost whatever usefulness it might once have had—present-day architecture wastes precious resources and causes great losses of life. The act of building, whether it involves giant hydroelectric dams or a single small home, is an act of land destruction. Buildings destroy land for as long as

they stand. The true costs of things like houses and parking lots and sewers and incinerators, not to mention quarries, mines and timberlands, are paid, year after year after year, in such currency as topsoil, trees, plants, insects, birds and mammals—currency we've hardly begun to recognize.

SELF-SERVING MOTIVES

All too often our motives for building ecologically are unashamedly self-serving; we talk a lot these days about the ecological revolution, about new lifestyles and about new priorities, but we tend to think more in terms of new versions of old mistakes—safer detergents, cleaner-burning automobile engines, that sort of thing. I don't see that very many of us are committed to any real change, and that's a shame because those who have tried to simplify their lives—and I count myself as only one of the most timid among them—seem to share a unanimous and very genuine conviction that real riches—the kind Thoreau was talking about—increase in direct proportion to the simplicity of one's life.

But what about the rest of us, those of us who forget or have never learned, as Thoreau did, how

UNDERGROUND DESIGNS

125

to see the miracle in everything around us? Are we doomed to want bigger and more comfortable houses forever? The answer might have been yes if nature hadn't suddenly begun putting on the brakes. Nature has now warned us that there's nothing but a dead end down the luxury/growth road. We've got to simplify; there's no other choice.

REMODEL OR BUILD?

But what to do—buy and remodel or build from scratch? Fortunately, it isn't an either/or kind of choice. Both new houses and rehabs can be built successfully if the construction work and the planned-for facilities are based on life principles. Look: here's what wild land does:

    (1) creates pure air
    (2) stores rain water
    (3) creates pure water
    (4) produces its own food
    (5) creates rich soil
    (6) uses solar energy
    (7) stores solar energy
    (8) creates silence
    (9) consumes its own wastes
    (10) maintains itself
    (11) matches nature's pace
    (12) provides wildlife habitat
    (13) provides human habitat
    (14) moderates climate and weather, and
    (15) is beautiful.

When *we* build we do just the opposite. We fail on every point of this life-list except one, number 13. We always seem to provide that one no matter what the cost.

Everything we build, from front porch to city, is ecological failure. It isn't based on reverence for life. Not yet, anyway; but the next architecture will be. Its central rule will be this: improve the ecological health of the land in question or don't build there. It's as simple as that. Then we'll be forced to revive the most devastated of sites first. Slums. Worn-out farmland. Strip mines. Old parking lots. They're the kinds of places on which to build, and the results can be glorious; a whole new kind of architecture.

NEW PRIORITIES

As our priorities shift back toward life, this is what we'll do to house the new person:

A. *Build Reverently:* Use the most abundant of local materials or those whose production seems to cause the least amount of damage to the land.

B. *Build Simply:* Use as few materials as possible. But don't skimp on important things like first-class waterproofing and super-extra-double insulation. They'll repay your efforts for the life of the building, which brings up the need to

C. *Build Permanently:* Instant domes and throwaway buildings sound appealing, but their use gives nature no time to heal the wounds of construction before the next round starts. Each time we move, uproot, repave, regrade, or break ground, we tear the fragile fabric of life on the land, a fabric which may have taken decades or even centuries to develop. We must build hundred-year and two-hundred-year buildings. Inside, their occupancy and decor can be changed whenever necessary, but for God's sake, no more ticky-tacky! It's too expensive.

D. *Build Naturally:* Make sure your project and its site, whether one house is involved or a thousand, do most of the following:

    (1) create pure air (trees, shrubs, vines, grasses, wildflowers)
    (2) create pure water (slow runoff, mulch, percolation)
    (3) store rain water (ponding, percolation)
    (4) produce their own food (this is a tough one!)
    (5) create rich soil (mulch, compost)
    (6) use solar energy (if you solve this one you'll get three Nobel prizes, and mankind will move ahead three giant steps)
    (7) store solar energy (another Nobel for this one, too)
    (8) create silence (dense plantings, sound insulations)
    (9) consume their own wastes (live organically)
    (10) maintain themselves (permanent materials, earth cover, good waterproofing)
    (11) match nature's pace (build permanently)
    (12) provide wildlife habitat (dense plantings, berries, shelters)
    (13) provide human habitat (a foregone conclusion)
    (14) moderate climate and weather (windbreaks, dense groves of native plants)
    (15) are beautiful (if you achieve the first 14, this one will be automatic).

E. *Build Personally:* We lost a precious thing when we became the only animals incapable of building their own nests. The miracle that is a brick will be forever lost to the man who never lays one. Build with your hands as much as you can; you'll never regret it.   ■

PART III

TOM BENDER

TOM BENDER

128

BEYOND APPROPRIATE TECHNOLOGY

The movement which has its roots in the ideas presented here is rapidly maturing. It is, in fact, so widespread and diverse that calling it a movement seems somehow inadequate. It is now a whole body of experience and knowledge which supports new approaches to business, architecture, energy, economics, law, health, community development, communications and agriculture.

People working in these areas have focussed their attention on widely different problems and have chosen a wide range of settings in which to work. The most visible government/institutionalized programs are the National Center for Appropriate Technology (68), California's Office of Appropriate Technology (70), and the Department of Energy's new Office of Small Scale Technology (71). Lessons about looking for the solution in the problem box are being learned here.

The Alternative Energy Resources Organization (AERO) (64) in Montana, ACORN (63), Rain (48), Farallones Institute (66), the Institute for Local Self-Reliance (67), and many others continue to develop—some with outside help and some without. They have been extraordinarily successful at pushing these ideas into the mainstream.

But progress has not been without setbacks. There are still existing institutional barriers which work against new ideas. The sheer power and momentum of corporations are taking them into the solar manufacturing business, which they now threaten to dominate. Legislation being passed in dozens of states in an attempt to stimulate new technologies actually inhibits low-cost, innovative, owner-built techniques. The public still believes that resource shortages are a hoax. Not the least of the difficulties is people seeking personal and political power by making appropriate technologies a national goal. All these developments bear careful, critical watching.

The issue of using low-cost technologies for low-income groups is only beginning to be addressed. Some projects have been successful and some have not—all for a variety of reasons. Karl Hess, in Flight from Freedom, *talks about the problems of introducing "living lightly" into the ghetto where people have not yet had a chance to live with material wealth.*

A major strength of appropriate technology is its fundamentally grassroots approach. Yet, as Tom Bender notes in Why Big Business Loves A.T., *the control of these tools may be lifted out of our hands if we concern ourselves only with technical improvements and neglect the politics of their manufacture and use. Corporations are waiting for the bugs to be worked out of alternative systems before they move in to mass produce them in a way that would shatter their decentralizing, humanizing potential. The real work now must be to insure that the means of making and using new technologies rests solidly in local communities.*

In the transition to more resource conserving lifestyles, sacrifices will be made—many of which will turn out to the good. Margaret Mead's essay explores the quality and direction of that transition. It is important, she points out, not to confuse the pangs of readjustment with the pangs of deprivation. We have adjusted before when we needed to—willingly—and we can adjust now.

If the work we are doing is to be successful, it is crucial that we be aware of the pitfalls and weaknesses while keeping our dreams clearly before us. As the final essays in this book show, there are increasingly hopeful developments in communities across the country which both endorse the feasibility of these tools and, in fact, make them ever more viable alternatives for large numbers of people. David Morris looks at some of the emerging legal and economic trends which are making local, democratic control of neighborhoods and cities possible. Lee Johnson explores the future of the most hopeful energy trend—community energy systems and integral urban neighborhoods. Gil Friend evaluates new forms of agriculture, taking a look at new accomplishments and areas which need more work. He concludes that more attention must be focussed on changing the structure in which an organic, ecologically sound agriculture can prosper. And lastly, Tom Bender sums up the spiritual, personal, economic and political realities and futures of these technologies.

Together these new pieces provide a vision of the possibilities for a coming era. □

Although there are numerous examples of technology projects, there are very few places where a neighborhood has attempted to deal with the interconnected issues of neighborhood power, self-reliance and intermediate technologies. Examples of this in low income communities are even more rare. In Flight from Freedom, *first published in the October issue of* Quest/77 (47), *Karl Hess describes what happens when a community of diverse backgrounds and interests makes an attempt to organize and improve itself. More than anything, this account is a reminder that none of the issues with which we are dealing are simple, nor are the problems easily solved.*

FLIGHT FROM FREEDOM
Memories of a noble experiment

By Karl Hess

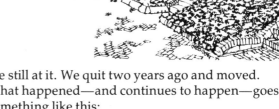

ADAMS-MORGAN is a small country afloat in a great city. It is a 70-block neighborhood in the center— almost the exact center—of Washington, D.C. The population is 58 percent black, 18 percent white, 22 percent Latin, with the remainder mostly Middle Eastern. It is a neighborhood in transition; as a small country, it's in decline.

For a while, during a rash and wonderful tilt at making itself a true community, Adams-Morgan was a fascinating culture in which to live. More recently it has become a prime target for real estate speculators, a bullish market well beyond the means of the people who first made it a good neighborhood. Its nature is slowly changing to chic—from workshops to boutiques, from bars to cocktail lounges, from a heady whirl with community government to an enclave of town houses with barred windows, and residents whose concerns are global rather than local. If the present trend continues, the neighborhood will eventually disappear, and Adams-Morgan will be just another Washington address. A fancy one.

For almost five years, Therese Machotka and I worked with hundreds of people in the neighborhood striving for an entirely different future. Some

are still at it. We quit two years ago and moved. What happened—and continues to happen—goes something like this:

I spent my childhood in the neighborhood, got my first haircut in a local barbershop that is now a locksmith's shop, kissed my first girl in the part of Rock Creek Park that borders Adams-Morgan. I had my first fistfight under the bridge that carries fashionable Connecticut Avenue safely past the northwest edge of the neighborhood, and I attended one of the two schools from which Adams-Morgan derives its name.

After 40 years or so I came back. What had been comfortably middle-class had become very lower-class, a shambles about to become a slum. But it was cheap—and it tolerated hippies both socially and economically. This was the mid-sixties. The hippies who moved in were mostly exiles, mostly useless, and the neighborhood slipped down another notch. Venereal disease went up. Panhandling became the local growth industry. Welfare blacks and zonked-out whites began to drink Ripple together and curse the dark night of colonialism, oppression, and shortages of good hash and sturdy H.

By the late sixties something began to stir in the debris. In the manner of the opening scene in *2001*, some dazed hippie got sick of the faucet dripping or the VW van not running—some minor calamity—and, wonder of wonders, got straight long enough to FIX IT. Perhaps the change began more subtly, a complex process susceptible to sociological jargon. My own experience is that it was fairly simple and direct. Somebody had to do something. Someone did. It worked. And the world changed a little. Odd jobs became more satisfying than panhandling. How-to books began sliding into shelves alongside the mystics and the revolutionaries. And something very important began to happen to the residential warrens in which the stoned citizens had compartmented themselves.

By the end of the sixties, there were probably 60 to 75 functioning communes in the neighborhood, and a burst of productive energy emanated from them. A worker-managed grocery store opened and thrived as a place to find good prices and good-natured advice about nutrition. Then a second one opened. A local newspaper popped up, reporting neighborhood news. Then a second one. A record store. Several bookstores. Crafts people, from potters to auto mechanics, began hawking their wares from community billboards, tree posters, street corners. Musicians rented a storefront and began nightly sessions of jazz, rock, country, classical. Several graphic arts shops opened. A community credit union was started. A community government proclaimed itself, called a meeting and actually got off the ground.

The government arose, like everything else, from the rubble of failures immediately past. Until then, Adams-Morgan's neighborhood organizations had merely presented resolutions *to* the city government. None had dared to *be* a government.

At the first meeting to discuss something new, the pioneers were young, white products of the counterculture who proposed a breakthrough—not just another civic group, but a town meeting. In homely, practical terms, they argued that a town meeting would bring neighborhood people together to state their problems, discuss solutions and decide what they could actually do themselves—without futile complaining to the sluggish, federally controlled city bureaucracy.

The thing was called AMO (Adams-Morgan Organization). At its first meeting, someone argued that the streets were dirty. Someone else suggested a clean-up day. The meeting agreed. Signs were mimeographed on a church duplicator, paper was donated by a man working in a print shop. The neighborhood was saturated with the information that AMO members (then only about 300) were going to sweep down the main street over the weekend. About 200 people actually got out and swept. Nearly all of the neighborhood's 40,000 residents heard about it. People began to perceive AMO as an organization more interested in doing than in talking. By the time we left, the membership exceeded 3,000.

The town meetings—or AMO Assemblies, as they were called—were the most exciting political experiences I have ever had. After sampling a participatory democracy, I would never trade it for a merely representative one. Still, there was a small problem: not everyone accepted the difference between participation and representation.

The counterculture people were looking for a new way to make social decisions without one group exploiting another. They were inspired by the idea that town meeting participants not only made decisions but also carried them out, rather than getting someone else to do it. As a result, counterculture types made up at least half of every Adams-Morgan meeting.

Blacks had a clearly different view. They accepted the rhetoric of participation, but they were mainly concerned with representation. In that neighborhood, at least, they were not interested in changing the way social decisions were made. What they wanted was the power to make those decisions—power *in* the system, not power to change it. Whites who do not understand this can make mistakes in assessing black-white alliances.

Therese and I had our own special interest. Although we were among the town meeting's most active participants, we were preoccupied with science and technology—not in the abstract, but in terms of helping the neighborhood. We had long worked with the Institute of Policy Studies on a project aimed at identifying ways of life and social agreements in which people are full participants. As advocates of political decentralization, we focused our IPS work on learning how to put science and technology at the service of a free society in which people are truly responsible for their actions. Such freedom, we felt, requires a system of mutual support, enabling people to join democratically in deploying the tools of everyday life and production. Our laboratory was our neighborhood—Adams-Morgan.

We began weekly talks with a couple of engineers

we had met through the peace movement, some craft friends and students. Others joined and soon tired of talking. Therese and I coaxed a neighborhood clinic, operated by Children's Hospital, into letting us have unused space in the warehouse building it rented. Therese agreed to put most of her salary as an editor into buying equipment and paying stipends for work. Our talk group became a project that we called Community Technology.

Our weekly meetings continued, sometimes crowded with 40 or 50 visitors, as an information-sharing process. A young physicist with superb general mechanical skills came on full time for a subsistence share of the money Therese made available. I covered our living expenses by writing, welding, selling metal sculpture and occasionally lecturing. Our experiments began.

Food, it seemed to us, was the place to start. What could be more basic? Also, the idea of developing food production in a ghetto neighborhood seemed as stern a test of our general propositions as could be imagined. There's no land for growing food in a city. If there is any open space, it's too much trouble. Yet the land problem was easily solved. Food grows, not in an abstraction called land, but in a reality called someplace-nutritious-to-put-down-roots. Space for this reality need be only that—space. We located a lot of it.

First, the rooftops. Most of the neighborhood consists of three-story row houses and apartment buildings. Nearly all the roofs are flat. On very strong roofs, organic soil can be spread, or boxed, for growing vegetables. Therese and I grew such a garden. Less sturdy roofs accommodated the lighter demands of hydroponic growing—the cultivation of plants in tanks of liquid nutrients or in nutrient-soaked sand. Friends who began a companion enterprise called the Institute for Local Self-Reliance, still a prospering activity, operated a hydroponic garden with storybook success and wildly bountiful crops. They also managed to fill virtually the entire neighborhood's demand for bean sprouts from a single basement facility.

More traditionally, we worked with kids in the neighborhood to establish regular gardens in vacant lots and in any backyard space that people wanted to make available. The entire back lot of our warehouse was covered with dirt that we begged from local excavators and converted into a community garden. Also, using the vegetable wastes from several local grocery stores, leaves from suburban lawns, and horse manure from a park police stable, we maintained about 90 feet of compost pits behind the warehouse.

To supplement the vegetable crop we looked around for a suitable meat animal. Cows were out. Too big. Rabbits didn't make it. Too cuddly. Chickens wouldn't do. Too noisy. How about fish?

One of our group, an organic chemist, was experienced in trout farming and suggested that we work up some high-density indoor tanks for raising that fancy fish.

Plywood tanks sheathed in fiberglass were built by Jeff Woodside, our resident physicist and jack-of-all-trades, together with his immensely energetic friend Esther Siegal, our chemist Fernwood Mitchell and Therese. To recirculate the water, they used pumps from discarded washing machines. To handle the fish waste, they contrived filters made of boxes filled with calcite chips (the standard marble chips sold in garden supply stores) and a couple of cups of ordinary soil. The soil provided nitrifying bacteria that fed on the ammonia in the fish waste.

The bacteria kept the water clean. The pumps and some well-placed baffles kept the tank water moving in a strong current. The fish (which we first reared from eggs in ordinary aquarium tanks) swam strongly, ate heartily of the commercial feed that we first used as a convenience, and grew as fast as fish in streams. In fact, they converted their feed to flesh at the rate of one ounce of fish per two ounces of food. This made them just as efficient as chickens and about 500 percent more efficient than beef cattle. Our installation, neatly tailored to urban

basements, produced five pounds of fish per cubic foot of water. A typical basement in the neighborhood could produce about three tons annually at costs well below grocery store prices.

Other members of our group tried other projects:

A young stonemason began experimenting with small, completely self-contained bacteriological toilets. He had fair success demonstrating that any neighborhood can unhook itself from conventional sewer systems, with their inefficiency and pollution.

A marine engineer built a very effective solar cooker that tracked the sun automatically and cost under $300. Using energy collected by an outside mirror, in the shape of a three-foot-long trough, the device heated an indoor hot plate up to 400 degrees Fahrenheit.

The group generally began discussing the design of a shopping cart that could be built in the neighborhood; a self-powered platform that would handle most of the neighborhood's heavy moving chores; a neighborhood chemical factory to make household cleansers, disinfectants, insecticides, even aspirin; and a neighborhood methanol plant to convert local garbage into a fuel roughly similar to gasoline.

We sought a grant from the National Science Foundation to start a science center where local people could study the natural science of the neighborhood, possible tools and techniques, and the appropriate role of science and technology. The NSF sent a sociologist to look us over and turned our application down cold. We did not fit the government definition of a neighborhood self-help program. Even at NSF such programs are aimed at enhancing a group's ability to get more welfare—not to mold its own economic future.

Government programs aim at giving money to poor people. Our hope was that knowledge would, in the long run, be more useful, even provide more money, and eventually strike at the systemic causes of poverty. Government believes that poverty is ac-

THE Alternative School, PA

tually a lack of money. We felt, and continue to feel, that poverty is actually a lack of both skill and the self-esteem that comes from being able to take charge of one's life and work.

It will not be denied that ours was and remains a middle-class attitude, quite classical and thoroughly Western. It stands opposed to the elitist notion of mandarins caring for benighted peasantry—an attitude that prevails today in various modern trappings, among them enlightened capitalism, state socialism and welfare statism.

But so much for big notions. Reality was something else.

At Assembly meetings, reports of our work were always greeted with applause and great enthusiasm We were a showcase bunch of wizards doing wonderful, far-out things. Our appeals for neighbors to join us in the work, to help improve the fish farm, to move the gardens along, to experiment with new ways of growing, to start stores and even factories based on our skills and tools, got choruses of right ons—and no participants.

Instead, the Assembly began to emphasize direct appeals to government agencies and foundations for grants. More and more people wanted to make complaints about landlord abuses, not make plans to buy them out.

At meeting after meeting, for instance, we discussed the idea of pooling money to establish neighborhood ownership of key properties, to provide homes for the evicted, to set new patterns of ownership for a new kind of neighborhood. Plenty of right ons. No cash. Was there any cash? Of course. Even people on welfare have disposable incomes. The pool of money needed to buy our neighborhood would have been relatively modest, the weekly equivalent of a carton of cigarettes or a bottle of whiskey from each member of the Assembly. Of course, it would have meant sacrifice. Some of us have little enough pleasure, and a smoke or a drink is to be treasured beyond all the promises of paradise.

There were, in fact, jobs aplenty in the neighborhood. The District of Columbia government, sternly charged by federal authorities with making the streets safe for visiting dignitaries, including congressmen and bureaucrats, had decided that bribery was the best tool available for getting young people off the streets. The District funded programs through which teenagers could draw a minimum of $1.75 an hour for the exertion of signing in each morning and signing out each afternoon. There is no convincing evidence that this did anything to halt incipient criminality. It seemed to me that it accomplished a great deal more in terms of separating young people from the possibilities of self-reliance. It anchored them more firmly to habitual dependence on unearned incomes and, thus, on the people who dispense them—be they welfare bureaucrats or those less willing providers, the victims of larceny.

A question began nagging a lot of our work and discussion: Was our vision of neighborhood self-help crumbling along racial lines? Are blacks particularly disabled when it comes to seeking alternatives to welfare programs?

The Adams-Morgan neighborhood, like Washington overall, is certainly black. The people who seemed to talk most about, and do the least in support of, our group's proposals were black. Young whites seemed to respond more to skill- and production-centered activities. Those are solidly middle-class values out of a primarily European culture. Blacks have been the victims rather than the beneficiaries of both the values and the culture.

Black think black, as they continually say. In Adams-Morgan, at least, black has come to mean poor and oppressed. Black demands have come to mean black reparations: to be given something, rather than seeking the chance to do something.

Everyone in our largely white group abhorred racial discrimination. Some went further and supported the implicit strategy for social redress through reparation rather than community renewal.

My problems and doubts began with my conclusion (shared by Therese) that such a strategy was not only useless; it was unjust, crippling and ethically debilitating.

As I nursed such doubts, the Community Technology work began to seem quite different to me. I no longer thought it truly relevant to what was happening in the neighborhood. The hope that people would want to fashion new lives, based upon new knowledge and new skills, seemed now very romantic and very wrong. Desirable, yes, but hopeless.

There was another problem. Crime. It too fell along racial lines.

At one AMO meeting, a young white man reported a particularly vicious holdup, beating and rape that had occured at a communal house. Before any discussion could get under way, he was asked the color of the victims. White. He was asked the color of the attackers. Black. With blacks in the majority, that particular meeting simply moved on to another topic.

This typified a particularly destructive, if fashionable, impulse among both blacks and whites to dismiss all discussion of crime as oppressively racist despite the fact that blacks are the principal victims of black crime.

Another major victim of unchecked (because unmentionable) crime has been the AMO organization itself. Keeping a typewriter available in the office has always required a rigorous and none-too-successful exercise in security. Money needed for important things is eaten up in replacing routine equipment, rather than being devoted to attacking the roots of the problem—the need for self-reliance instead of welfare services.

Given a situation of rising, undiscussed crime, the inevitable results is neighborhood deterioration —and, ironically, the arrival of more affluent but less neighborly residents. These people can afford better security and don't mind living in a small fortress so long as the address is fashionable. This is now happening in Adams-Morgan.

Could neighborhood people have coped with crime? I certainly think so. It would mean first coping with their own children, facing them down, creating families that would absorb their energies and deserve their loyalties. Not easy. Not likely. And particularly not likely when parents are opiated by a welfare existence, and where schools are simply disciplinarian baby-sitters, offering young people no creative alternative to violence as the way to get out, to get up, to get even.

It was the growing crime and violence that finally drove Therese and me from Adams-Morgan. After being robbed every 60 days or so, Therese began to feel terrified at night in the neighborhood. We both resented the constant loss of things, particularly since our income was roughly at the poverty level. Being larger and male, I had not felt the terror. After a time, Therese felt it sharply enough to leave. We moved to West Virginia. Having been raised in rural Wisconsin, Therese found the West Virginia hills immediately hospitable in ways the city's streets had never been. I still love the city, but I have experienced a homecoming in these hills that is richer than I anticipated.

I still believe from my experience with the small workings of Community Technology that science can thrive in a neighborhood. Ordinary people can get together to discuss physical principles just as well as they can discuss abstruse political principles in the fashion of young radicals and young conservatives.

But the culture of poverty is not easily weaned from the illusory ''cure'' of program handouts. If this culture of poverty is to be broken in any black neighborhood, I am convinced that it must be broken by black people, not by starry-eyed whites talking soul patter. But even the most adventurous blacks will have trouble. Coming from his Chicago base, Jesse Jackson lectured black Washington teenagers on the need to learn skills rather than gripe endlessly about feeling oppressed. He was virtually run out of town for his effort. Washington's blacks are still infected by the notion that their problems are mainly political and will disappear when black power surpasses white power. Many are less interested in a better world than in a black one.

When the Assembly focused on local problem solving, rather than conventional constituency politics, it was greatly effective. Shy people spoke out. Seemingly hopeless people sparked to new life. Now all this is fading as the old idea of representation begins to recover the ground lost to the experiment in community participation. The Assembly has become more a bandstand for aspiring politicians than a forum for people.

A similar malady afflicts some of the worker-managed enterprises that brought the neighborhood to life in the first place. The malady is ideology. For several years, the workers in those enterprises toiled hard at being useful to the neighborhood and good friends to each other. Now several of the key groups have begun to work equally hard at becoming friends, not of people, but of history. They spend hours, behind closed doors, thrashing out the correct line on this or that remote political issue or revolutionary posture. They have forfeited real social power in Adams-Morgan and become figments of history floating in clouds of rhetoric and theory.

Blacks, by and large, have moved wholly into the rat race of conventional politics and foundation grantsmanship. And upper-middle-class *arrivistes*, both black and white, share no concern for the neighborhood beyond the recent trendiness of its address.

Some of the original spirit persists, however. The people at the Institute for Local Self-Reliance go on doing what they can, but more frequently this entails reaching into other communities—which

would be fine, except that it stunts the growth of neighborhood resources. With Therese gone and rent no longer provided, our Community Technology warehouse has been turned into a soap factory operated by Jeff and Esther, stalwarts of our group from the start. They make a living at it and they try to teach a few kids in the neighborhood how to read. They grin and bear the annual vandalism of the gardens by kids who think that vegetables are underclass food, while TV snacks, beer and dope are the fare of real operators.

The weekly meetings have ended. Information is now swapped by phone and mail. But almost everyone who was involved in the effort retains faith that it was a right thing to do, and that someday the memory of it will finally inspire neighborhood people to take their culture, their lives and their productive possibilities wholly into their own hands. Such a neighborhood will not change the world overnight, as in the fervid dreams of the young revolutionaries. But it will make a small part of the world much better. ∎

At the other end of the spectrum from urban, low-income communities working towards autonomy and self-reliance are the corporations who see what can be gained from the production of smaller scale technologies. This essay points to the crucial distinction between any technology and appropriate technologies—the values which guide their development, production and use. It was first published in Rain *(48) in May, 1978, and is reprinted with permission.*

WHY BIG BUSINESS LOVES A.T.

By Tom Bender

THINGS SEEM TO BE moving along almost *too* well for a.t.—the ideas of community production, decentralization, local control of local situations, etc., seem to meet encouragement rather than resistance from government and business. The power companies are ominously quiet, sometimes even co-operative. Executives of large corporations came to listen to Schumacher in droves. *Small Is Beautiful* is a best seller. ERDA, AID and NSF have set up a.t. programs. The U.S. Congress is interested in a.t. Carter meets with Schumacher and Amory Lovins.

Yet these are not compatible bedfellows. What's happening? Why does Big Business love a.t.?

Ask them. Surprisingly, you're likely to find out, but also to have some fantasies about a.t. nicely shattered. Ann Becker and Carol Ulinski did a study a year or so ago on IRRI, the International Rice Research Institute in the Philippines (*Development Digest*, Jan. 1976). IRRI has developed a number of highly useful small-scale agricultural tools and machines adapted specifically to the Southeast Asian rice culture and has been working to encourage the production, marketing and use of such machines. They found out, in brief, that Marsteel, the dominant machinery manufacturer in the country, encouraged IRRI's efforts to get small businesses to adopt and produce their designs locally. Marsteel's logic was simple. Their marketing manager, Luis Bernas, freely admitted that they were quite willing to let someone else do the hard and risky work of developing designs and production, demonstrating, testing, overcoming local inertia and developing markets for new products. They were confident that if demand for a product did develop they could step in and gain dominance in the market through their economic and advertising power and ability to overwhelm, fairly or unfairly, the small producers. Big business also welcomes the use of appropriate technologies where they either give a boost to the economic well-being of an area or lower costs—in either case making more money available for the purchase of their products.

Sound at all familiar? Take another look at solar energy developments in the U.S. What's the pattern? You have a situation that clearly lends itself to decentralized, at-home application and local production and installation. You have individuals developing and refining the simple technology required and fighting the massive efforts of entrenched energy companies, financial institutions and government that have done their best to prevent the rapid conversion to solar energy. Now that

MEG DE MOLL

the public has begun to demand application of solar energy, you find the government (ERDA in this case) giving massive amounts of *our* tax monies to pay large corporations to reinvent these already proven technologies. What is at stake is not *inventing* the technology but paying the corporations to develop *their* capabilities to produce it and also to receive credit from the government for inventing it. So the government promotes and pays big business to take over a new field that is developing quite well without its "assistance."

The next step is in process now. It is easier and more convenient for business to let government legislate the successful small producers out of business rather than have to compete directly against them. How to do it? Set up "performance" standards tailored to the capabilities of large corporations. The corporate approach to solar has consistently been biased towards exotic "high-efficiency" systems—ones that maximize the energy collected per square foot of collector but which produce less energy per dollar of expenditure. They know they can't compete on whole-system performance, so they try to push the issues to specific subsystems that can (but shouldn't) be maximized. The result is that an

apparently innocent technical standard for thermodynamic efficiency clearly discriminates against simpler systems (homemade or local collectors, wood heat, passive solar construction) that are overall more effective and economical. Look at the federal and state standards being set up to determine what designs will qualify for tax credits, rebates or financing, and see what they really mean.

Curiously, no one has been speaking up against this. What has happened to the people developing solar energy over the last decade who lovingly espoused the vision of decentralized, do-it-yourself technology? Burned out? Bought out? Shoved aside?

Their silence has occurred in part because solar has been so new that its proponents have welcomed *any* means to get it developed, endorsed, accepted and applied, realizing on some level that solar must be accepted in concept before the question of How and Who can be dealt with. More importantly, development of solar energy has been tightly tied with undevelopment of nuclear energy, and to many people the corporate control of solar energy has seemed to be an acceptable price to pay—better they make solar collectors than reactors.

But the solar/nuclear tussle is being won by the

inherently better wisdom, logic and economics of solar—not by corporate control of the industry. The question now is do we really need or wish to pay that price or if we can still avoid doing so.

In both the above examples there are several distinct issues to be dealt with. The use of solar energy or more effective agricultural tools is one issue. *How* those technologies are produced and used is a quite separate issue. Despite its name, a.t. is not dominantly a technological thing. It is the doing of things at the proper scale in the proper ways to the ends that create a society and a world of which we are happy to be a part.

Once the feasibility of the necessary technology is shown, the real task of a.t. becomes a political one —dealing with the changing of the institutions that tie us to certain forms of doing things. That is the stage we need to move into now. It has hit home here in Oregon with compost toilets recently. Legislation permitting use of compost toilets was passed in the last legislative session and the Plumbing Advisory Board given the responsibility of drawing up regulations for their use. Everyone relaxed. Then draft regulations appeared, limiting use of compost toilets to areas where sewers or septic tanks were not available. This would effectively turn compost toilets into a tool for real estate development rather than an economical and ecological alternative to sewers. The technology is not enough without the appropriate institutional forms affecting its use.

The efforts of most a.t. people so far have been on proving its feasibility, but we will reach a dead end unless we turn from promoting harmless things like solar collectors and compost toilets to dealing with the real issues of institutional power and control that allow or prevent things from happening on more local and controllable levels. No community industries can succeed in a sustainable way, for example, until the role of advertising in centralizing economic activity and power has successfully been dealt with. The same is true of finance and unregulated federal taxation.

Our ability to deal successfully with many of these issues is greater than we may think. Our visions and our own capabilities have gained credibility time and time again in issue after issue—the nuclear/solar/conservation tug of war, in Vietnam, about pesticides—you name it.

In contrast, our country has built up a horrible record of lies, deceit and wrongdoing through the lack of vision and integrity of our decision-makers —both in public office and private actions. We damn other countries for violation of human rights but have violated human rights both here and abroad since our country's beginning, as badly as other countries do. Our government trains terrorists, troops and torturers for repressive dictators in other countries. The activities of the CIA and FBI in this country against Native Americans, Blacks, anti-war activists and others is finally coming to light and is scarcely less shocking. And both probably pall when the full nature of our "legal" economic exploitation of people is understood.

Our country desperately needs people involved in all levels of decision-making whose vision and integrity can be trusted. Who do you trust? The corporate leaders whose deliberate mistruths fill volumes from the advertising in every newspaper to the Congressional Record? The mercenary bureaucrats who drag down any designated path with a paycheck at the end regardless of where their hearts want to go? Or the silent academics who have failed to speak out on any of the real issues we face until their students have beaten the answers into their heads? Or is it the people who stand up bravely and stubbornly against the tide until the wisdom of their perceptions is finally understood. People echo whoever feeds them. Feed yourself.

It's time to work through the politics of changes— from inside and outside government; at home, in our neighborhoods and in Washington, D.C. And we need to remember that power *does* corrupt. It needs to be shared, localized and taken control of locally ourselves. Get to it. ■

There is no getting around the fact that lowering energy consumption, conserving resources, and bringing production and decision-making back to the local level involves extensive changes. The most fundamental of these will be in life-style, which many argue are next to impossible—the last, most enduring barrier to any real transitions. At the same time, we are told that wasting energy is the moral equivalent of war. Here, Margaret Mead, long an astute observer of our culture, clarifies the necessity and direction of this moral imperative in light of the innate ability of human beings to adapt.

CAN AMERICANS CONVERT TO LOWER ENERGY USE?

By Margaret Mead

ON THE WALLS of American factories and offices during World War II, a proud piece of self-confident bombast hung, proclaiming in various idioms our confidence in ourselves, in our resources, our ingenuity and our capacity to do what needed to be done. "The possible we do at once," it read, "the impossible will take a little bit longer." And we did. Overnight, when war came, out of the dissidences and hesitations, the mistaken old world loyalties, the raucous insistence of the left and right that there was no choice except that between Fascism and Communism, the country pulled itself together, got to work and converted into a wartime economy. Whole new towns sprang up as great volunteer mechanisms were set in motion. This was accomplished in spite of the growing disillusionment with war, with the engines of war and the propaganda apparatuses which underwrote the merchants of death. We did this with a wry quirk in our smiles, without returning to the simple naivete of the first World War, without the self righteousness of opposing sides, each of whom prayed to the same God to destroy the other. The theme song of the Americans in the Second World War became: "Praise the Lord and pass the ammunition." (I had to search long and hard and in vain in Shakespeare and the Bible to find its moral predecessor: "Trust God and keep your powder dry.")*

But the wartime sense of purpose was only a respite. It was a brief return to the possibilities of total commitment, in the corroding polarization which was settling over Europe and the United States, as there appeared to be no good choices, only a choice among greater and lesser evils; between Fascism, where the trains would run on time but at the price of aggression, oppression and death, or Communism at the cost of ruthless sacrifices of other millions. Political philosophies, born of the exploitative realities of the industrial revolution, struggled with each other for dominance, while waves of new moralities and new technologies spread over the world. (No new political philosophy or political inventions have since appeared that reflect, as those in the 19th century, the emerging technologies—of the steam engine, coal and iron—and new capacities to exploit the non-renewable resources of the earth.)

---

*American folk version of a remark attributed to Oliver Cromwell.

We are today letting ourselves be maneuvered into trying to answer the wrong questions. We are being asked and we are asking ourselves whether we can go back, can give up, can shrink our ambitions: Can we return to the outmoded life-styles of a time when we knew no better, to the stone pestle and mortar, the hand pump, the horse and buggy? The advocates of "small is beautiful" continue to present their case against a monster, while the proponents of the assorted vested interests of the world continue to proclaim: "Nuclear and proud of it," falsely identifying the realizations of our deepest moralities with unnecessary and dangerous nuclear constructions which endanger the future of the planet.

If you ask any people with a shred of pride and determination (vestige of the older, simple patriot-isms that led men to stand and die on the borders of a loved country) whether they are willing to go backwards, there will be many who will say NO. Sweden phrased a retreat from imperialism as just that, a retreat. In spite of the high ideals for a better human life-style, somehow the mainsprings are slack. Britain, weary from standing on the outposts of the world, liquidated the British Empire and failed to attain a lesser, but morally responsible role. At the other extreme, the attempt to find a better life for the majority of the people is also phrased as "giving up" by the Communist parties around the world. Only by "giving up" the diamonds and rubies of the rajahs, the great estates of the ruling classes, the comfortable stuffiness of the land-narrow horizons of the middle class, the raucous-ness of the trade unionist who is concerned with control of his tools, the liberalism that believes that progress is possible; only so, are we told, can the world escape disaster.

The overdependence upon oil, and accident of the burgeoning of modern technology, has given a curi-ous archaic twist to this out-of-date reasoning. The old economic doctrines of the 19th century were based on the idea of shortages; that there was not enough wealth to go around and that only by dis-tributing it so that more people each got less could we ever hope to attain a semblance of social justice. Then, during and immediately after World War II, we began to glimpse the possibilities of the elec-tronic revolution, when pattern and design would replace brute steam and the internal combustion engine, when humankind's new understanding of the natural world could indeed produce a new material basis for human life, a reliance upon inex-haustible sources of energy, upon recycling rather than exhausting resources. From the beginning of the electronic revolution, humankind has, for the first time, the possibilities of providing more for all, instead of less for some, if there is to be more for the many.

Glimpses of the extraordinary possibilities which faced the human race gave temporary form to the emerging moralities of the late 1940s. For the first time in human history, we could seriously look forward to a world from which the older universal forms of suffering—hunger and thirst and cold and unnecessary premature death—could be banished. In the first wave of confidence, colonialism in its old forms was swept away, giant steam mills were en-visioned for the least technically developed coun-tries. Enthusiasts spoke of the speed with which new industrializations could take place—in 30 years or in 15 years—without the pain and the horrors of the industrial revolution, without England's "dark Satanic mills."

That vision was correct. A new order was within our grasp, but it was not the 19th century order of giant machines, endless premiums on bigness of scale and on the initial exploitation of cheap labor. Our realization of the tremendous possibilities of the electronic revolution only permeated the spread of the old technologies in the form of providing new possibilities for eliminating expensive labor and substituting automation. There was a simple paradigm: manual work had been the mark of ex-ploitation, of serfdom, slavery and indignity since the beginning of a socially stratified society. Drudgery, the shackling of both men and women to the endless daily grind which accompanied human-kind's enbracing a settled life, was associated with manual labor, with the routines which, in hunting and gathering societies, were the lot of women and children. Anything that freed anybody from exert-ing physical effort—walking instead of riding, pro-pelling on wheels instead of carrying, artificially fabricating instead of painstakingly shaping with skilled hands—was voted as good, whether it was for the housewife in rural America with water pumped into her kitchen, or the New Guinean, recently freed from fear of his headhunting neigh-bors, who substituted an outboard engine for pad-dling against the currents of his mighty rivers or on the open seas when the wind failed.

We recognize today, as we mull over the disasters which have accompanied the spread of the new technologies—only made possible by the electronic revolution, under the old slogans of the indignity of physical labor, of the human desire to reduce painful, back-breaking, premature aging—has continued to capture the imagination of the people in both the industrialized countries and the newly industrializing countries. Everywhere in the world technologies that promise greater speed and greater convenience prevail over older forms of hand labor and transportation. But they are powered by forms of economic and political power that belonged to the old fossil fuel economies, that favored the locations with coal and iron and continued to require access to cheap labor.

We now realize that the world has reached a critical stage of interdependence, has attained horrendous powers of destruction and lacks the political organization to deal with these new conditions. The critical nature of our present condition is true, but the way we are phrasing it is misleading. When we ask: IS it possible to reverse the trend towards resource-intensive life-styles, all those who say ''No'' are assuming that this trend has been good, that the post-World War II world has been a rewarding world to live in. All of those who say, ''We must'' assume that the present style is wholly bad and that there is no answer except to *go back* —back to the land, back to hand tools, back to all labor-intensive activities, back to the hand pump, the wormy vegetables, the rotten eggs, the sickly children who had had no vitamins from their winter diets, back to a death rate where so many babies died that almost all men and women had to spend their lives producing and feeding children, back to limiting our abilities to travel, to explore, to probe the mysteries of the universe.

I do not believe that any society can purposefully take a step backwards without flying in the face of all that we know about the evolution of human societies. But I do believe that the world took a wrong turn after World War II. We started on a path of enormous possibilities but with the accident of cheap oil—a historical accident which was not written in any blueprint of evolution, which was one of those historical accidents that can, nevertheless, re-rail a civilization and today may indeed destroy our interdependent work—the need to develop new

forms of inexhaustible energy disappeared. Transportation costs became no consideration so production of food, fuel, renewable but fragile resources became concentrated in those parts of the world where the greatest profit could be made. Instead of the logic of the marketplace as a means of powering a capitalistic system and the prudence of political power as a way of modulating political adventurism in socialist regimes, cheap fuel became a determining factor.

We have almost wrecked the world with massive deforestation, pollution of streams and oceans, depletion of the soil and decimation of marine life, depopulation of the countryside and the construction of huge unmanageable cities, location of nuclear weapons, and now production of nuclear power all over the earth, producing a population explosion in which the dependent young—in the industrializing countries—and the too independent elderly—in the industrialized countries—threaten to overwhelm our capacity to care for them, disenfranchising millions of children of their right to be reared in families that give them love and security, saddling future generations with a burden of nuclear waste from which there is no escape. All this is true. This is what has resulted from a combination of old outworn political systems and new unbridled, untested, poorly understood technologies.

But the answer isn't just to reverse a trend. The answer is to get control of this accidental runaway in our interdependent world system. The present day world does, for the fortunate, have many freedoms from burdensome drudgery, from the people of Bali who can ride on busses instead of walking two days to a ceremony, to the busy, overworked housewife who can unwrap a packet of frozen peas instead of having to shell them. But we will have to take a look at what these busses in Bali or frozen peas in the United States (from a monocrop planting of peas a thousand miles away) cost; they may be lacking in some essential mineral or vitamin, may contain dangerous additives, whose manufacture may have disrupted some balanced market gardening economy and which, in some previously diverse single-crop culture, may be destroying the natural vegetation or bird life of a whole area. The trends that need reversal are the headlong pursuits for narrow monetary gain by the producers, the unthinking enjoyment of the consumer, the applications of the new technologies. The old balances which could be attained in small states, in local market economies, and in a limited amount of distant trading no longer work, and we need to find ways to attain new balances. In the pursuit of these new balances, new kinds of safeguards, new legislation and new institutions, there will be many disruptions, as there have been during the half century when the present conditions were being established. But there is no need to confuse the pangs of readjustment with a permanent state of deprivation.

We can, I think, compare the kinds of temporary hardships and readjustments that have in the past been demanded in wartime with some of the demands that will have to be made of the American public if we are to take the lead in helping to establish a safer, more viable world interdependent system. But before we use the figure of speech of war, I think we need to clarify the way in which war can possibly be used to invoke appropriate peacetime behavior. President Carter has called for a "moral equivalent of war" and for sacrifice, and listeners have identified this call with deprivation and loss. Is this necessary? When the idea of a "moral equivalent of war" was originally introduced, it included the need for channeling what were presumed to be humankind's instincts of aggression, instincts that had led to conflict and war throughout human history. It was in this sense that the Olympic Games have been promoted—but have very doubtfully promoted peace—as a way in which these presumed instincts could be limited and still given expression.

I believe it would be better to think of the aspect of war for which we somehow need to find a moral equivalent as that behavior, found in its fullest expression in wartime, which leads individuals to make any sacrifice, to endure hardship, torture, mutilation and death for some overriding value which they attach to their country as they defend it. This is the spirit which has led conquered peoples to fight on for decades or centuries, which has sustained men when the odds were hopeless, which had made women willing to rear sons who would die as their fathers had died "for the ashes of their fathers and the temples of their gods." It is not the human tendency to attach obstacles to their goals and to respond to frustration with aggression

that we need to cherish and re-channel into less lethal forms. It is the human capacity to *defend* the nest, the homeland, the young, the unborn and the unconceived that has sustained human beings as we have evolved from members of tiny bands to the citizens of great nation states. Although the sacrifices that Americans have been asked to make have been far less than those demanded of the citizens of other nations, to the American mother who has lost a son who was one of hundreds of thousands rather than one of twenty million Russians, the loss is nevertheless absolute. Paradoxically, deprivations of the affluent in wartime or in depressions always seem greater than the deprivations of the poor. Those Americans who never got over the Depression of the 1930s were often those who had to give up one of two cars, not those who sold apples on street corners and could glow with joy when they had a bowl of soup. It is the well-fed who suffer conspicuously when near starvation hits, although it is those on the edge of survival who may die with the imposition of just one more deprivation.

Great or small, the loss of sons or husbands in battle, or the loss of comfortable living quarters, the need to import resources from overseas, or the limited freedom of movement from gas rationing, the sacrifices and deprivations of wartime have been conceived, since the invention of warfare itself, as terminable. We did not even speak of real warfare until the invention of truce and peace had also been made. War as distinguished from head hunting, or blood feuds, or piracy, is a special state in which a political group can transform its relationship to other groups so that those who were neighbors, trading partners, are transformed overnight into those whom it is a virtue to kill. In warfare, murder becomes patriotism; those who were once thought of as human as ourselves (be they Germans or Japanese or North Vietnamese) are transformed into less than human beings, who are outside the protections that human culture provides for all those who have been thought of as our "fellow men," women and children.

These transformations which can be accomplished by a declaration of war also include the capacity for reversal; just as former allies can become enemies, so also they can become allies again. The invention of war—and peace—was a step towards the creation of larger and larger social units; from the band to the cluster of villages, from the tiny chiefdom to the modern nation state. Without the steps provided by centuries of warfare, evolution would not have taken the form it has.

The aspects of warfare to which we have to turn when we want equivalents are the sacrifices which human beings are willing to make to defend their political identity—including their land, people, the unborn, their religious faith, their language, their political institutions—and the way in which human beings conceive of those sacrifices as terminable. Wars can be won or lost; someday there will be peace, and those things for which one has been fighting and suffering and sacrificing will be again within one's grasp. Someday the lights will go on again, and "normal" life will return. It may well be that the definition of a cause as hopeless and no longer worth fighting for and the realization of what constitutes normal life are very closely related. Perhaps those people who have invested the most in particular material things, whether they be land or paved streets and sewers, may give up the more easily because they are less committed to immaterial or unsubstantial values—like language, or freedom of religion, or freedom of movement. This may explain the apparent greater readiness to surrender, to compromise, to make terms with the enemy of the Belgians after World War II, for example, than the determination of groups like the Jews, Armenians, Gypsies and Greeks to struggle on during hundreds of years of foreign rule.

It may explain also the apparent greed of the common man once the war is over, his concentration on rebuilding his home, on the feasts he will give and the vacations he will take. It may also explain an even greater danger; the stubborn, unyielding conservatism of those who have survived and who want everything to be as it was before, as indeed what they fought and suffered for. When periods of social disorganization assume many of the characteristics of wartime, as in revolutions or the Black Death in Europe, or the Great Depression of the 1930s, they produce some of the extremes of, on the one hand, a desire for improvement over the pre-war conditions and, on the other hand, a stubborn clinging to old forms. Both are explicable, both are predictable, and both need to be taken into account in any social planning.

So, when President Carter calls for sacrifices, such as those made in wartime, it is not an impossible request for Americans to respond to, as they began to respond to the Energy Crisis. Americans, once convinced that some sacrifice is necessary and that everyone will be asked to make the same sacrifices, make such sacrifices with fair alacrity and cheerfulness, as we accepted what rationing we had during World War II. But it must be seen as necessary, fair, and imposed on everyone. In World War II, sugar rationing was accepted, but gas rationing, which was so diversified that every category seemed to be some kind of abuse, was not. If the measures proposed during the beginning of the Energy Crisis of the 1970s had been imposed uniformly on the entire nation, instead of the oil states planning to benefit from the hardships of the states without oil or the truckers imposing their separate demands, we would have acquiesced and would be much further along in coping with the end of cheap oil. We muffed it and we set back our national capacity to adjust.

When, however, the demand for wartime-like sacrifices is coupled with threats of a future unending state of shrunken and miserable life, then the wartime analogy, the demand for patriotic national sacrifice, rings hollow. This is just what happened with the publication of *Limits to Growth* in 1972 and the spreading pessimistic acceptance of the Marxian definition of the economics of the modern world. American businessmen, instead of rising to the challenges of their entrepreneurship as they did in World War II, accepted the gloomy prognostication that the days of the free enterprise system were numbered and that they had better get whatever they could at once. They took their capital abroad for quick returns, failed to build up their plants at home, and in their single-minded, short-sighted pursuit of monetary profit, contributed grievously to our present state of American overdependence on the rest of the world. Americans, who respond to opinion polls by saying that, they think, despite their distrust of every American institution, America is a good country to live in, are speaking out of this short-sighted gloom: it's a good country now, now, when we have all that we have, but it isn't going to stay that way. As the cars are getting worse each year, and soon we may not have any, many people are buying the biggest one they can now. Inflation also contributes to the pessimistic view of the future, and to a dangerous, apparently very selfish and short-sighted view of the present; whether it is the dictator of a small banana republic lining his pockets, or of a woman on the verge of retirement buying a large, expensive new car, or the leaders of the Soviet Union at the end of World War II, realizing what rebuilding their devastated country meant, preached "No rest in our time," thus commiting themselves to substituting coercion and fear for hope.

The first need in the United States, as well as in all of the industrialized countries, is for conservation—cutting down on our reckless consumption of oil, reducing pollution, stopping our consumption of artificial fertilizer, reducing our waste of food and waste of scarce materials. These are the only measures open to us in the short run. These can be compared to wartime sacrifices, and the president can call for a moral equivalent of wartime commitment. But it must be clear, at the same time, that *we do this as a means of survival, because there is a better life ahead*.

Changing the technological style of the last 150 years and particularly of the last 50 years need not involve any loss, and it can involve a great many gains for the mass of the people. Our life-style has involved despoiling the land and our natural resources on an unprecedented scale, exploiting distant peoples, constructing exploitative economic and political systems. True, it has also involved hope, hope that mass manufacture and large-scale enterprises would give humankind a better material base than they have ever had. But as long as our future political life has been postulated on fossil fuels, on large-scale centralization of production and power, it has also offered us a grim picture of a future in an over-populated world where there was not enough to go around.

The contemporary scientific revolution has changed all this. We now have the prospect of utilizing inexhaustible resources, of controlling population growth by humane and medical measures rather than by war and famine. We have the capacity to make anything that we need without despoiling the earth. Furthermore, we have the possibility of eliminating all of the disastrous by-products of the industrial revolution—mass poverty, separation of human beings from the natural world and participation in basic production, fragmentation of social, political and intellectual life, monstrous contrasts between the rich and the poor. The things that we

will lose when we build a way of life based on the knowledge we now have—knowledge we didn't have even 25 years ago—are things we can well do without. We can dispense with the senseless and wasteful production of food that is guaranteed not to nourish, with the hours and hours of busing and commuting of children and adults, with the separation of our daily lives from the kind of exercise that keeps human beings healthy. We will neither find it necessary to manufacture and pay for "no-cal" foods, nor buy expensive mock bicycles to ride in the dining room, nor perform expensive operations to cut out part of our stomachs which should never have been enlarged. The world at present, and particularly the world in the affluent countries, is a patchwork, band-aid world, desperately trying to undo what shouldn't have been done in the first place. On the one hand, states pass laws against non-returnable bottles or the use of saccharin, baby clothes made of inflammable materials, or soaps that are non-biodegradable, and on the other hand, in desperate reaction against all change, medical professionals legislate against the use of diet as curative, and educators try to return to methods that worked 50 years ago.

Appropriate technology does not mean making water pipes of clay or bamboo because these were once, or are now, appropriate, durable and economical somewhere in the world. It does mean planning so that whatever kinds of pipes and wires and lasers we find best to use—best in the widest sense, best for the world and for posterity—can be grouped in utility corridors so that the cities of the future won't have to be torn up every week. It does not mean abolishing all the discoveries of nuclear power; it does mean stopping the use of nuclear power for electricity, for light and space heating and local transportation, stopping, as Barry Commoner has said, using bulldozers to kill flies. It does not mean the end of mass production, but it does mean reestablishing some relationship between the product and its use so that industries will not deliberately choose to make things that are dangerous, because short-term profits will be higher. It need not mean abolishing private enterprise, but it does mean becoming much clearer about what enterprises are appropriately private, what should be left to citizenship efforts, what to profit making and the search for bureaucratic and political power. It does not mean getting rid of political motivations; politicians have to be elected if they are to serve good ends as well as to serve narrow and destructive ends. It does not mean that we will not have to have planning on a grand scale and sometimes, as in the uses of sea and the air, on a global scale. But it does mean that we will need to sort out which operations should be planned at what level, which operations should be performed at a distance and which closer to home. It does not mean destroying thin tropical soil by exhausting it with crops that can be better grown on the open plains of the temperate zone. It also does not mean using five times as much energy to produce one unit of energy output—as we do in our present United States agribusiness. It does mean a vast readjustment in the way the world has recently been going.

Can we do it? Of course we can. We can change our direction now just as we have changed it before. We can convert to low energy, greater settlement density, localization of manufacture, suitable transportation, just as we converted to our present high energy, dehumanizing, destruction-bound present way of life. And, fortunately, conversion pays. It paid Detroit to build tanks instead of cars in World War II; it paid Detroit to return to cars; it paid United States industry when TV came in. It would pay to build the machinery needed for appropriate technology around the world—simplified machinery, and machinery and equipment needed by the developing world, suitable housing, freight cars, public transportation. We did it before and we could do it again.

The possible we do at once, the impossible takes a little longer. ∎

The past ten years have seen a maturing of the alternative agriculture movement: organic food production is becoming an increasingly respected science, alternative marketing networks reach across states and community gardens flourish in all our cities. Yet these encouraging developments are only part of a wider picture which Gil Friend explores here. A co-founder of the Institute for Local Self-Reliance (62), Gil now splits his time between the Office of Appropriate Technology (70), pursuing a master's degree in agricultural ecology at Antioch College West, and working at Farallones Institute (66) to develop a Center for Sustainable Agriculture.

NURTURING A RESPONSIBLE AGRICULTURE

By Gil Friend

WE HAVE BEEN WITNESS this century to one of the largest mass migrations in history. It was invisible to most of us, perhaps because it happened in bits and pieces. There were rarely crowds filling the streets. More politely spoken of as a population shift, this migration was the massive movement of Americans from the farms to the cities. The information about it is usually presented in the form of percentages which deaden the impact of its scale. A shift in farm population from 9.4% of total in 1959 to 4.2% in 1975 says one sort of thing. A shift of almost eight million people off the farm in less than 20 years says quite another.[1] For the entire century, the numbers are even more striking. And the pattern is the same in the developing world as well.

This transition has also represented an enormous cultural shift—one that has been taken almost for granted. When noted, it is with pride. Wendell Berry quotes a former Deputy Assistant Secretary of Agriculture, James E. Bostic, Jr.: "But just stop for a minute and think what it means to live in a land where 95% of the people can be freed from the drudgery of preparing their own food."[2] That figure disguises the larger number of people involved in the off-farm portions of the food system; its accuracy will be examined in detail below. But perhaps we

should do just what Bostic suggests, consider what it means, or rather what it would mean otherwise. For one significant feature of appropriate technology in agriculture is that it may facilitate a different distribution of people between city and country, and between agriculture and other enterprises.

A more appropriate agricultural technology will not in itself drive such changes. Like any other technology, it is a means to an end, not a determinant of that end. But a broader range of technologies may in fact provide a broader range of choice. With an expanded range of efficient technical options to choose from, people may come to realize that social and industrial evolution are not scientifically dependent on a process of natural selection of technology. Technology is often developed in response to explicitly or tacitly selected goals. And goals can be changed.

In relation to agriculture, appropriate technology has usually called to mind (and is often called) an "intermediate" technology, midway between the simple digging stick of Third World subsistence agriculture and the sophisticated, energy-intensive machinery of U.S. agribusiness. If we were to focus exclusively on hardware, then these intermediate

146

technologies would be our main interest. Certainly an intermediate scale of technology is needed if small farmers are to be economically viable in the U.S., and even more so if poor farmers in the developing world are to increase their productivity without being displaced from their land.

But appropriate technology is not all that is needed for an alternative agriculture to flourish in the midst of what has mistakenly been called the most productive agriculture in the world. An agriculture which reflects the values usually implied by a.t. advocates—commonly emphasis on resource conservation and minimal environmental degradation, occasionally concern over popular control of technology and resources—will require not only a different technology but different patterns of organization of production, distribution and administration and a different paradigm of how to manage biological systems. Because agriculture is dealing with conscious management and manipulation of living systems, any applications of technology that are to be appropriate must also include an appropriate conceptual approach to that management.

The most important single issue in assessing the future of "alternative" agriculture, the role of appropriate technology in that future and, in fact, the significance and value of any of a wide range of issues touching on things agricultural, is the matter of sustainability. In a society that has stressed short-term maximization rather than long-term optimization of benefit—and benefit to the individual actor rather than benefit to the surrounding community—as the prime criteria for action, it is not surprising that sustainability is not only of little importance in management decisions, but is hardly ever *mentioned* in that context.

It is understandable that the interest in and the means for this "peak production" agriculture found their fullest expression in a country that developed its agriculture as it developed its frontier and its industry. Facing the apparently unlimited potential of westward expansion and industrial development, it must have been easy to ignore the more sobering potential of soil exhaustion. Scientific understanding of mineral depletion by crops was not yet fully developed in the 19th century, so such attitudes at that time cannot really be faulted.

But these attitudes were not only the result of scientific ignorance. That ignorance existed in a cultural and economic milieu which believed in and demanded unlimited expansion of production and extensive exploitation of land and labor. Scientific awareness has matured considerably in the last century, but the surviving economic imperatives and cultural priorities are still determining factors in agricultural practice and policy.

Underlying any form of social organization, however, any set of social and economic priorities, is a biological reality that must be reckoned with. While economic and political and cultural factors drive the specific choices exercised in the time span of the fiscal year, the electoral term and even the human lifetime, it is the biological world that we are based in, and the physical world that *it* is based in, that set the larger parameters for our more familiar choices. Try as we may to sidestep these biological limits, it can only be a temporary escape. Inexorably, we are reminded of the reality of those limits by loss of topsoil, rising energy prices, deteriorating food quality, biocide addiction, decline of rural culture, and growing waste management problems. We are compelled to consider an approach to agriculture that will produce food, fiber and income in a way that is sustainable, maintaining the productive capability of agricultural lands and the vitality of rural community life long into the future.

Some have suggested that we may look back on the 1970s as a period of transition for agriculture, a key point when, if the trends of modern agriculture were not changed, at least the trend of the trend began to shift. Agribusiness maintains its presence, family farms still fall to creditors and developers, and food prices continue to rise while food quality diminishes. Yet, these analysts maintain, the momentum is no longer with those phenomena, but with the growing movements of biological agriculture, natural foods, anti-corporate awareness, food cooperatives, community gardens and related efforts.

How realistic is this vision? Will these years form the basis of real, lasting change in the way we grow food, or will they represent one more hopeful, but temporary, ripple on the face of a more depressing history? For there have been hopeful ripples before —co-op movements in earlier depressions, farmers' strikes in other periods of cost/price squeeze. (Indeed, financial instability seems to be more the rule than the exception for the American farmer—a meager lot for the producers and stewards of a society's most basic wealth.) Yet, these movements have always evaporated with improving conditions, only to reappear with the next downturn. Or they have failed to consolidate their gains into a durable political power able to affect the root conditions of their troubles. Have we learned enough from these earlier efforts to do a better job of making our gains and keeping them?

One of the greatest values of alternatives, even those which have not endured, is that they confirm the feasibility of another way. They serve as a reminder that the common wisdom is not the only wisdom, but merely one option among several that has emerged, or been selected, in response to particular conditions. Thus, for example, food cooperatives and regional co-op federations serve as a reminder that efficient food distribution can be organized without the driving motive of private profit.

The most durable single effect, though, of alternative agricultural activity of the past decades is likely to be the lessons of ecological agriculture. Just a few years ago, the agriculture establishment could ignore organic farmers as a handful of kooks. Former Secretary of Agriculture Earl Butz was able to make the preposterous claim, five years ago, that "before we go back to an organic agriculture in this country, somebody must decide which 50 million Americans we are going to let starve or go hungry."

The ongoing studies of large conventional and organic farms in the U.S. Corn Belt by Dr. William Lockeretz and others have shown that organic agriculture cannot only be productive on a large scale, but can be economically competitive as well.[3] The enormous energy savings realized by the organic farmers in the Lockeretz studies could not fail to catch the attention of agricultural policy makers during a period of high energy prices. The efforts of regional federations like Tilth in the Northwest, the Natural Organic Farmers Association in New England, and the Agricultural Marketing Project in Appalachia and the South are showing, albeit less visibly, that small farmers can produce ecologically and efficiently if suitable support structures are available. As another example, recent research on some California crops has shown that farmers can earn as much or more per acre by substituting integrated pest management (IPM) programs for exclusive reliance on pesticides.[4]

In response to such documentation, and to the all-important pressures of rising energy prices and tightening environmental regulations, the agricultural establishment has become much more open to options it had ignored, or even ridiculed, a few years ago. (Interestingly, it is the older scientists who remember an agriculture without massive chemical inputs and huge machines who are more open to the organic option than the more conservative, younger scientists who were educated during the golden age of pesticides in the '50s and '60s.) Funding opportunities for research related to organic agriculture have mushroomed. In early 1978, the Farmers Home Administration removed its long-standing restriction on credit to organic farmers, who had formerly been considered a bad risk. Many banks in the cotton belt no longer provide credit for pesticides, recognizing the economic as well as environmental disadvantages of relying solely on toxic chemicals. Direct marketing programs, and extension programs targeted at small farms, are receiving support in a growing number of states.

As important as these new signs are, however, they are in themselves small islands in a big stream. The poor state of the U.S. food system has never been the result of a dearth of examples of another way—the Amish, for example, have been quietly with us for many years—but rather of the economic and political conditions under which it has developed. As exciting as recent alternative efforts have been, they are struggling to develop and take hold in the midst of the same conditions that have plagued farmers and consumers for decades. Organic farmers may be slightly better off than their "conventional" neighbors, by virtue of lower input costs, but they are still vulnerable to the mad oscillations of the commodity markets.

American agriculture can readily absorb a new body of technique without altering the existing relationships of production, trade and control. The ability to change while remaining the same is one of the remarkable strengths of American capitalism. As long as organic farmers hold to the dream of being successful small entrepreneurs in an economy dominated by consolidated capital, they will maintain that vulnerability.

The food cooperative movement has been subject to a similar short-sightedness. Co-ops have been able to supply a specialized clientele with some savings, but cannot compete favorably with the giant food corporations for supplies or credit. Much of their success to date has been based on voluntary self-exploitation by dedicated workers. But such a strategy cannot last long, especially as the workers get older, raise children and think about future security. Living lightly and increased self-reliance may reduce this problem, but only in degree, not in kind.

For the transition to a sustainable agriculture to take hold, we need more than changes in degree. We need especially to understand that the present pathologies of the U.S. food system are not the result of ignorance or accident alone, but of responses to real conditions in the physical and economic worlds. Unless we can begin to exert leverage on those conditions themselves, farmers, consumers and corporations will make choices that may be rational for themselves in the short term, but wrong for society in the long term. Such a perspective can eliminate a lot of hand wringing and can help clarify steps necessary for durable change. As we come to understand that the American pattern of extensive agriculture—large holdings and abuse of soil fertility—was a ''rational'' response to cheap land and energy, high labor costs and control of transportation and markets by big capital, we can see that mere regulatory restrictions on farm scale are not enough. There must also be programs to create the positive conditions for agriculture of another sort to flourish.

This may seem like a call for an uncomfortable change of gears for many alternative agriculture/food co-op/appropriate technology people. Perhaps so. It is certainly a call for people to stretch themselves beyond their personal interest in organic farming or food cooperatives or solar collector design to understand more about the economic conditions those efforts face. Organic farmers would not plant a seed they know a great deal about in a soil they know nothing about. Similarly, the vision of a sustainable agriculture will not take hold if its proponents work only on the ecological and technical issues that may initially have attracted their attention. We must also understand and change the economic and political factors that ultimately affect everything else we attempt.

WHERE ARE WE NOW?

What is the present state of the collection of phenomena we are calling the alternative agriculture movement? Beyond the general points made above, it may be useful to outline several strengths and weaknesses of the movement as it stands today and where future efforts need to be directed.

Five Strong Points (what has worked):

• Development of the techniques of ecological agriculture

• Growth of urban agriculture

• Development of new marketing patterns

• Application of appropriate technology in agriculture

• Effects of personal and social attitudes

Five Weak Links (what needs more work):

• Farm prices

• Marketing

• Preservation of prime agricultural land

• Development of appropriately scaled machinery

• Direction of government research, market and extension programs

THE STRONG POINTS: WHAT HAS WORKED

Ecological Agriculture

The techniques of ecological agriculture, once ridiculed by agricultural scientists, have not only proven their worth, but have shown themselves to be a coherent and thorough approach to farm management. It has its share of folktales, but a sound body of experience and theory as well. There are many variants of organic technique, at best keyed to regional soil and climate differences, at worst to naive beliefs in ''one true way.'' But there is a com-

149

mon body of principles underlying them all. Briefly stated, they include:

• understanding and managing the farm as a complex ecosystem, with particular attention to the biological community of the soil;

• focussing on long-term optimization rather than short-term maximization of the productive capability of the farm system;

• maintaining a diversified farm—rather than a monoculture—where the various living components of that ecosystem can perform many complementary functions that would otherwise have to be provided by the farmer;

• using organic wastes, green manures, rock powders and other slow-release nutrient sources to close nutrient cycles and maintain soil fertility to the extent possible; and

• avoiding synthetic chemical inputs.

Common understanding of organic agriculture among some experts as well as lay people has all too often stressed only these last two points—the use of organic wastes and avoidance of synthetic chemicals—when in fact something far more complex, basic and thorough is meant.

More and more farmers—a surprising number, in view of the limited agriculture extension information available for the transition to organic techniques —have managed to make that transition and are thriving with it, as much as any farmers are thriving these days. For many it was very much a choice of necessity rather than belief. Deteriorating animal health under intensive chemical techniques is the most common reason farmers give for going organic. There is as yet little scientific research to confirm the results, but farmers themselves commonly report dramatic turnarounds in animal health, fertility and milk yields after changing their soil management approach. What research has been done to date has focused on economics and has found the economic viability of organic farms, on a variety of scales, competitive with that of their chemical-using neighbors.

The success of these pioneering farms is the best advertising organic agriculture can have. Farmers are, wisely, a conservative breed, slow to adopt new practices (though probably quicker these days

with the enticements of advertisements and salespeople). Visible demonstration of successful organic agriculture right next door may be the best persuasion to try it. Sometimes, though the reaction is continued skepticism and hostility. Some organic farmers have reported that their neighbors are convinced they spray pesticides on their organic crops under cover of darkness.

In recognition of these successes, and the need for more visible working examples, a number of demonstration and training centers have been or are being established in several regions of the country. Some examples: Graham Center in North Carolina, Women in Agriculture in Massachusetts, Farallones Institute in Northern California, and no doubt others. The Family Farm Development Bill of 1978 even calls on USDA to establish such centers in each state.

Even more rapid acceptance in the mainstream is being realized by integrated pest management (IPM), which may be thought of as a subcategory of ecological agriculture, though it is often practiced without the entire body of principles mentioned above. IPM is a perfect example of appropriate technology in agriculture. It is labor- and knowledge-intensive rather than capital-intensive. It is appropriate to the biological reality of the farm rather than the economic reality of the pesticide industry. Its value is even recognized by such large growers as Gallo in California, which may be a mixed blessing. On the one hand, they are able to achieve major reductions in pesticide use (IPM does not eliminate pesticide use, but rather uses it as a complement to more primary biological and cultural techniques), which reduces accumulation of toxic substances in the environment, farmworkers and produce. On the other hand, it is a reminder that organic techniques in themselves will not change the organization of agriculture. It is far more likely that, as has happened many times before in our history, the amoeba of corporate agriculture will absorb innovation and make it its own, with little change in structure and without any more equitable distribution of wealth and income.

Urban Agriculture

Urban agriculture has bloomed in the 1970s. Vacant lots in cities across the country have been put into cultivation—a reminder of the days not too many decades ago when considerable vegetable production and even livestock were common even in large eastern cities. Recent Gallup polls estimate that half the families in the country now grow some of their own food, and that millions more would if they could get access to land. Several million dollars of new USDA money has awakened sleepy urban cooperative extension offices in a number of cities, and CETA funding enables many cities to maintain full-time support staffs for community garden programs. Agriculture in the urban fringe could come to represent a significant food source with even greater economic and employment potential. Market gardens in or near urban areas are a common feature of many European cities, where special tax exemptions permit their economic viability. Similar developments in U.S. urban areas are under study, but are still moving slowly.

One obvious benefit of all this activity, and for some the major attraction, is the economic advantages of local production. For many, the benefit of fresher, higher quality produce is a more significant attraction. But there are social effects which are important as well. Gardens provide a basis for interaction and cooperation that are all too often lacking in urban neighborhoods. Often they are the first step in what becomes an unfolding series of community activities.

More subtle, but perhaps more important in the long term, is the gradually increased awareness of the production aspects of the food system that urban people can develop through these activities.

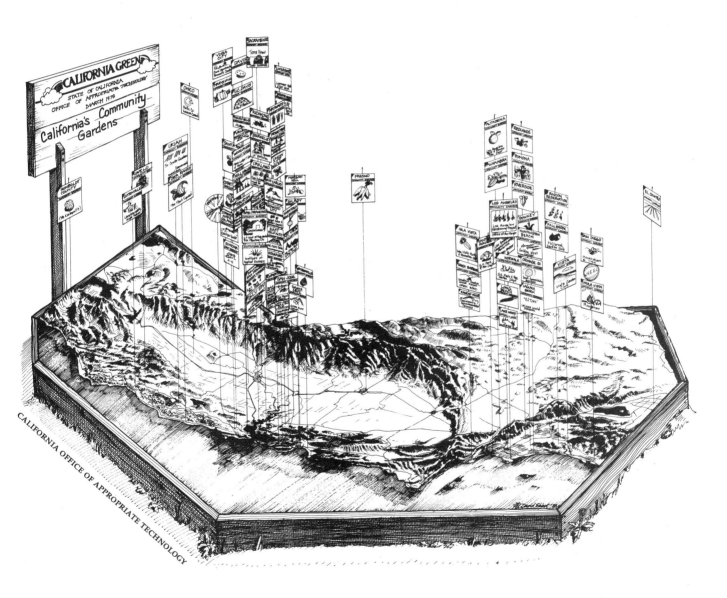

151

It is a step towards understanding more about the environmental issues affecting agriculture—which city people, with their legislative majority, can affect profoundly—and toward comprehending some of the complexities that face the farmer. It is just a step, but one that it is almost impossible to make for someone whose only contact with the U.S food system is through the grocery store.

Marketing

The flowering of alternative marketing structures likewise has both economic and cultural benefits, and is one of the strengths of the alternative agriculture movements. Consumer cooperatives have provided low cost food and varieties of food often not carried by local supermarkets. They have flourished in recent years as they have before during hard times. Buying clubs have spread out of student areas into low income neighborhoods and even into the suburbs.

Scale and organization range from large member-owned firms like Consumers Cooperative of Berkeley (California) to neighborhood buying clubs with volunteer labor and rotating chores. Between these are the workers' collectives and stores which have mixed paid and free labor. All but the large structures (which are organized like any other business) have provided valuable experience in cooperative work and management.

Of greater concern to farmers, many of these businesses have been able to provide stable markets to small local growers who previously had to accept less favorable contracts with large food distributors —when they could get those contracts at all. Thus co-ops, usually acting as regional federations, can obtain the foodstuffs they want, often grown to their specifications, while the farmers gain secure contracts, often with partial credit, and the producer-consumer relationship is brought a bit closer together.

Direct marketing by farmers, while lacking the cooperative framework of the consumer structures, may do even more to bridge the misunderstanding and conflict that have often separated farmers and consumers. Regular contact in the local market with the producer of the food one eats can help build mutual understanding and reduce at least some of the conflicts of interest that have made consumers and farmers feel like antagonists, rather than allies, in the face of the giant food industry corporations

which drain them both. This effect can probably be seen best in Europe, where weekly farmers markets are a common sight. At about the same time U.S. cattlemen and consumers were cursing each other during the beef boycott of 1973, the clients of one market garden near Stuttgart, when faced with the loss of their food source to a housing development, raised several hundred thousand dollars to preserve that land and food supply.

An additional step towards bridging this gap between town and country is the development of labor exchanges. The Wisconsin Brigade in the early 1970s helped place urban people on small farms during planting and harvest. The farmers got the labor, the workers got the experience, and both gained in the interaction. The best example of this approach is again outside the U.S. A group of Tokyo consumers had built up a successful buying club system, but decided that it didn't go far enough. So they bought and cleared a stretch of pine forest, installed about 8 of their 100 member families as permanent residents, and now run a farm that produces nearly all their food. Each urban member spends several weeks a year working on the farm, and rural members can rotate into the city as well.

Appropriate Technology

On a more strictly technical level, there have been important strides forward. Solar technologies are rapidly finding their way into agriculture. They are pioneered, quite often, by small farmers and back-to-the-landers, but are finding increasing acceptance on large holdings and are attracting all-important research money from the federal government.

Two important examples are solar crop dryers and "solar-heated" greenhouses. Solar crop dryers permit both a financial savings and reduced dependence on unreliable propane supplies. Solar-heated greenhouses can greatly cut the single largest expense of greenhouse operation—heating—thus enhancing the feasibility of controlled environment production, whether producing as an independent industry or as an extension of traditional farm growing patterns. Intensive greenhouse production may also prove to be especially suited to urban agriculture.

Another example is composting machinery. Composting is an important waste management option, often a necessary part of non-chemical fly manage-

152

ment for dairies and feedlots. Efficient turning of large compost piles has been difficult, with front end loaders not ideally suited for the task and the early compost turning machines too expensive for all but large operators. A new generation of specialized turning machines has brought the price down considerably, and new approaches, including a simple attachment to common manure spreaders and even simpler suction aeration techniques, promise to make fast, hygienic compost production even more economically attractive.

Attitudinal Changes

One of the great values of the processes under discussion here continues to be their effects on consciousness and culture. Less tangible than solar technologies or organic techniques, these effects nonetheless are a crucial part of the changes the appropriate technology/decentralist movements hint at. And they form the context within which the more specific and tangible aspects of political change can develop.

The effects move outward from personal activity and awareness to new social consciousness and perception. For some urban cliff dwellers, it may be no more than a matter of greater personal satisfaction from being closer to life in a community garden. Others, as discussed above, gain a greater understanding of the pieces and connections of the whole food chain, and perhaps of the farmers' perceptions of it. Many have taken the next step of learning about the underlying workings of domestic and international politics and economics that affect food and agricultural issues.

The power of such a range of levels is that people can begin with what moves them and meets their needs, and then become exposed to the other factors and activities—farmers markets to pesticide regulation to anti-imperialist organizing. Not all people will respond to all of it, but the connections are all there.

WEAK LINKS: WHAT NEEDS MORE WORK

For all the forward movement of the past ten or more years, the quest for a sustainable agriculture in the U.S. faces serious shortcomings. For the most part, these are not shortcomings unique to this movement, but problems that face agriculture as a whole. The qualities which have distinguished "alternative agriculture" from the mainstream have,

for the most part, not exempted it from the general crisis of agriculture.

Price Policy

The key problems, as they seem to always be for farmers, are economic. It is most evident from the cries of the striking farmers last year that farming just doesn't pay. To some extent, these producers are squeezed by the capital-intensive, chemically dependent style of agriculture they practice. To that extent, smaller, less heavily debt-ridden, less chemically dependent farmers may be somewhat better off.

But the problem of farm prices predated $50,000 tractors and goes deeper. As the authors of *Failing the People*, an important exposé of the land grant college system, note, "Liberty Hyde Bailey [Dean of Agriculture at Cornell] observed in 1914 that the profits from farming were being distributed increasingly in the cities and not in the rural areas, and that is even more apparent today."[5] Farmers have long been at a disadvantage in getting a fair exchange for their products, which many would argue represents the most basic production of wealth in the society. Without agricultural surplus, industrial civilization is simply not possible, but farmers have often not so much traded that surplus as had it extracted from them.

Yet American farmers have always imagined, in the words of Thorstein Veblen, "That they are still individually self-sufficient masterless men." They have perceived themselves as independent people operating in a free market economy, speculating in a land by extracting wealth as industrialists would extract the wealth produced by their workers. Michael Perelman observes that "Veblen's ironic essay [*The Independent Farmer*] pictures the farmer deluded by his fantastic vision of the world around him. All around him, bankers, railroad magnates and food processors profit from their 'effectual control of the market,' while the foolish farmer does little more than identify with the very people who are most adept at exploiting him. Someday, he too will share in the prosperity of the system, at least so he believes. The poor farmer turned a deaf ear to the warning that it is certain that all cannot become rich in this way. 'All cannot be shavers—some must be fleeced'."[6] Veblen wrote *The Independent Farmer* in 1923, and his criticism still largely applies. During hard times, farmers sometimes recognize the need to organize and occasionally understand their conflict with concentrated capital, but the lesson is generally forgotten as the crisis passes. The potential for cooperation in American agriculture has yet to be realized and may have great impact when it is.

Even for those whose primary concerns are environmental, the price issue must become critical. If the only way farmers can earn a living is by pushing the land beyond the limits of sustainability, they will have little choice but to do so. Parity may come to be seen as an ecological issue.

Federal support programs could themselves provide that link. Former Environmental Protection Agency analyst Armand Lepage and others have suggested linking price support levels to soil conservation practices in a way that would provide a concrete incentive for farmers to explore and adopt many of the approaches of ecological agriculture.[7]

An urban constituency for a sustainable agriculture will have to outgrow traditional demands for a "cheap food" policy for an ultimately wiser program of fair prices to producers and fair distribution of income among consumers. Public recognition of the hidden costs of "cheap food" will be a necessary step for such a policy shift.

Marketing (again)

Marketing appears here, as well as under "What Has Worked," because it has been an area of only partial success for the alternative agriculture movement. The efforts mentioned above are real and significant, but they have only touched a small percentage of consumers and farmers and are vulnerable young shoots themselves. To compete effectively with the present concentration in the food industry—if that is possible at all—will require greater coordination to achieve economies of scale, and a higher degree of vertical integration among producers, consumers and distributors. The problem, if such growth and coordination are possible, will be to ensure that these new units, though controlled by farmers, workers and consumers, do not become the same sort of self-maximizing units that present private corporations are, and that even the large producers co-ops have become.

The commodity market system is beyond the scope of this essay, but must be mentioned. A responsible agricultural system will need to evolve other mechanisms for distributing the risk of crop failure to replace the giant, occasionally rigged, always unequal (the house always wins . . .) rou-

lette game of commodity trading. A national crop insurance system has been suggested as one option, and could tie in with federal support programs.

Land Preservation

Preservation of prime agricultural land is another key issue which has not yet been adequately addressed. Each year a million acres of prime land are lost, often irretrievably, to urban development. In view of the carrot and stick many farmers face, with rising costs, land taxes and inadequate income on the one hand, and an eager real estate market on the other, it is no surprise that so many farmers finance their retirement by selling to a developer. To date, there have been only a few pilot efforts in the U.S.— Massachusetts and Suffolk County, New York, among them—which have tried to reduce this economic pressure by buying development rights and assessing land at its agricultural, rather than its development value. Vermont has experimented with steeply graduated taxes on land speculation profits. In many areas land trusts have been established to keep agricultural lands out of the speculative market. But the general inadequacy of programs across the country is shown by the continued diversion of prime land into housing subdivisions and shopping centers. Much more is needed, in a two-pronged approach to remove development pressure and to secure farmers adequate returns on their labor.

Agricultural Machinery

An appropriate agriculture must have appropriate machinery. For a long time, as Wendell Berry explains in *Horse Drawn Tools and the Doctrine of Labor Saving* (elsewhere in this volume), agricultural technology evolved to help farmers do their work better and more quickly. In recent decades, mechanization just helped do the job faster, while the quality of work often declined. In addition, tens of thousands of people have been put out of work by mechanization. The trend will continue in the coming decade as lettuce and brassicas (cabbage family—broccoli, brussels sprouts, etc.) join tomatoes and cotton as largely mechanized crops in California. Many of these displaced people ultimately become a drain on the urban social services, all in the name of "cheap food."

Further, mechanization in the past has helped

squeeze small farmers out of the market by catering to the needs of large growers. Land grant colleges such as University of California at Davis have spent public monies developing machinery scaled to the larger, not average, growers in their areas. The smaller farms were forced to, in the words of Earl Butz, "get big or get out" since they couldn't afford the machinery at their existing scale and couldn't compete with the big growers without it.

It is possible, though, to envision an agricultural technology which is scaled to smaller farms, is energy conserving, and is labor enhancing rather than labor replacing. One has only to travel in Europe or Japan to realize that such machinery exists, and that more can be developed even though it may not be in the interest of the large equipment manufacturers to do so. But that need not be a problem. The latest entry into the field of appropriate machinery is a mechanized, labor-enhancing transplanter for small, intensive vegetable operations, developed by a father and son team on an Oregon family farm.

Research and Extension

Finally, an area where there has been minimal success because there have been minimal efforts. The primary focus in the past decade has been on developing the physical presence of ecological agriculture, alternative distribution systems and corporate accountability organizations. Little if any effort has focussed on building institutional support for these efforts. This has not been a mistake. The real world is a much more realistic and satisfying arena than the federal bureaucracy, and the energies of these past years were probably best spent where they were. But with the new openness of the USDA, for example, it is advantageous to bring that structure to bear on enriching, supporting and disseminating information about these approaches.

In the area of research, the federal government can move rapidly and has begun to do so. A slower, and ultimately more important, process is the redirection and retraining of the local extension agents. Farmers rely on those agents for a great deal, but few extension agents can provide the sort of information needed by ecological farmers. Extension agents will have to learn more about these new/old approaches for the increasing numbers of farmers who are interested.

Some might argue that the Extension Service should be allowed to evolve in its own sweet time and that farmers should work instead with the private consulting services that are springing up to fill the information void. The Extension Service, they maintain, is not worth the effort at this time. Certainly the experience of European biological farmers with private farm advisors—who often provide certification and marketing services, and in at least one case, cooperative research programs, as well as farm management advice—supports the feasibility of such a private sector or cooperative approach. But the resources of the USDA and existing systems should not be ignored.

FUTURE PROSPECTS

Where are we going? What lies ahead for alternative, or ecological agriculture? It is probably a mixed future. The adoption of ecological techniques will proceed fitfully. Scientists, government planners and bureaucrats will probably be most open, with the agri-chemical industries fighting any potential business losses tooth and nail. The early signs of that resistance are the already strident industry voices attacking the as yet unreleased (as of this writing) California statewide environmental impact assessment of pesticide use. Farmers will fall somewhere in-between. Many will stay with the advice of the ads and salespeople and what they have been doing for so many years. Others will be so pressed by input costs, animal health or a neighbor's very interesting results that they may be willing to try something a bit unusual. Rising energy prices and tightening environmental regulation of pesticides and agricultural runoff (non-point pollution) should drive the long-term balance in the direction of more ecologically sustainable techniques.

The economic future is not as clear, however, for it depends less on new technical wisdom than on political power. And it requires a change from the usual exercise of politics which pits farmers pushing for higher prices against consumers pushing for lower prices, with the food processing industry looking for the biggest possible middle.

What, then, are the main tasks ahead for food and agriculture workers and activists? These tasks are in part evident in the preceding discussion, but several others can be identified. Foremost is for each of us to broaden our vision to encompass all the interacting components of agricultural issues, not just our pet interests. Economists and political organizers must understand ecological principles that provide the physical reality within which society operates. Likewise, organic agriculture proponents and appropriate technologists must no longer focus on merely a technical fix. Technological change does not evolve in a process of natural selection in an economic vacuum, but more often in response to powerful economic and political interests. Only by coming to understand those interests as well as the historical unfolding of our present patterns will we be able to create the context for the kind of technical and ecological practices we want.

Beyond understanding our roots, we will need to be able to sustain our dreams. That will take a higher degree of cooperation than has been common in American agriculture. We have had consumers' co-operatives with a fitful history of success. We have had durable producers' marketing co-ops which have themselves grown into giant corporations too big for their members to control and which have concentrated on maximizing their interests against all social costs. These co-ops have won important gains for farmers, but raise the question of how to build cooperation rather than self-maximization among co-ops.

Perhaps there is room in the spectrum of our agricultural activity for another level of cooperation —in the production process itself. Michael Perelman notes that Veblen understood that 'farming is team work', and goes on to comment that "cooperation, not competition is the key to efficiency in agriculture."[8] Yet, aside from memories of 19th century utopian communities, old-fashioned neighborliness and the recent smattering of small communal farms, the spirit of the individual still persists in American agriculture. Collectivist models like the Chinese commune or the Israeli *kibbutz* may be too far for most American to go. But cooperative approaches like the Israeli *moshav*, in which land is held privately and certain equipment and labor, as well as marketing, are shared, are reminiscent of the once common cooperation of rural America—barn raisings, harvesting teamwork, and a general spirit of shared endeavor that should not be so difficult to recapture. Recreating it in economic fact will not be easy, of course, but nurturing relations among people is a goal as important as nurturing the fertility of the land. Further, cooperative effort may be the only way that the many rural poor, and particularly the black poor of the deep South, can gain control of their land and succeed in agriculture.

Strengthening regional and local food economies is another key task for the alternative agriculture movement. Without ignoring the interconnectedness of regions, and of the U.S. to the rest of the world, we must re-establish ecological and economic health on the local level as a sound basis for health in the larger system.

Attention to regional food systems is probably strongest in New England. The local food economy of this once self-reliant region has been severely eroded, with 85% of most foodstuffs now imported. The potential benefits, both in lower food prices and in revitalization of the local economies, were first recognized by public interest and alternative agriculture groups and were actively promoted by the co-op federation. In recent years, state agencies have joined the campaign. Programs to encourage local producers and local markets are underway in New England as well as in other parts of the country. In a number of regions, researchers are beginning to examine not only food production potential, but also organic waste resources to explore the feasibility of a closed nutrient cycle, sustainable agriculture on a regional scale.

In ecological terms, more localized food economies, with the greatest possible proximity between food production and food consumption, could enhance the development of a sustainable agriculture. Reduced transportation costs would make recycling wastes onto near-urban agricultural land far more feasible. Energy would also be saved by reduced transportation of the food products themselves. The diversification that regional food self-reliance would have to mean could, though would not automatically, contribute to the diversification of the individual farm unit that ecological agriculture requires.

On the more immediate economic level, the two key steps are, first, reduction of the cost push on farmers (often reducible by a shift to ecological techniques and adoption of more appropriate technologies) and second, revitalization of local marketing opportunities. The latter is not merely a matter of opening farmers markets. In some areas where that has been done—parts of Vermont, for example— local demand is simply not great enough to absorb local production, and farmers have glutted the market in many crops. The scale of "regional" programs must honor such factors, as well as political and biogeographic boundaries. For the most part, such programs have been the province of state agencies or private groups. But the potential for action by municipal governments affecting both distribution and urban and near-urban production is considerable, though few cities have seriously explored that potential. Hartford, Connecticut, is one exception that bears watching.

If governments are to be moved to action, however, and, more important, if people are to support the redirection of the U.S. food system that all these activities imply, then "alternative" efforts must become mainstream interests and programs. They must become appealing beyond the growing, but still small, circle of eco-activists, natural food fanciers, back-to-the-landers, and a few born-again farmers. To do so they must not only capture the imaginations but also address and meet the real and perceived needs of the people for whom food is *just* something they eat or sell and not a symbol for so much more. Programs, therefore, must deliver concrete short-term benefit to people—and to more of them all the time.

Yet they must do so without settling for only that. The goal of the alternative agricultural movement, and the decentralist/anti-corporate movement of which it is part, is not (or should not be) merely cheap food for consumers and a better living for small farmers. It must also strive for an educated and organized citizenry that can work together to create the different social, economic and political patterns needed to sustain that better and more equitable life.

It is a difficult task. It is more difficult for its contradictions, which require us to consider long-term as well as short-term implications of our activities. Even then, the issues are not always clear. What is the delicate distinction between adoption and co-optation of a good idea by business or government? If ecological agriculture techniques benefit large corporate farmers as well as small family farmers, is that a good or a bad thing? Clearly, it has elements of both.

The very notion of "self-reliance"—among the most basic ideas of appropriate technology advocates—has its contradictory aspects. It is an empowering concept, returning productive power to the hands of "the people," and an important metaphor for political power, affirming that the roots of both forms of power lie with the people, not with the state or the corporation. Further, appropriate technologies, self-reliance and living lightly can make economic survival during hard times—which are likely to be more the rule than the exception during the coming decades—far more feasible by greatly reducing the cost of sustaining an equivalent quality of life. Its value

as economic self-defense can even extend to serving as a cushion that can help working people survive long strikes. Yet, while supporting such important activities, self-reliance can, paradoxically, permit greater exploitation. More self-reliant workers can maintain their standards of living on lower wages, thus permitting greater profits to their employers and reinforcing an economic structure that needs major overhaul. (This is not an abstract projection, but a technique that has been consciously used by such diverse operations as British mining firms in the 19th century colonial Africa and Henry Ford in 20th century America—both of whom gave workers' families land for food gardens.)

This does not mean, of course, that we should not continue to work for and support self-reliance in general and a sustainable agriculture in particular. But it does mean that such actions should be taken with full awareness of their contradictory nature, and with constant attention to whether and how short-term efforts contribute, educationally and materially, to longer term goals of not merely a sane agriculture, but a sane society. ∎

1. Michael Perelman, *Farming for Profit in a Hungry World: Capital and the Crisis in Agriculture*. Montclair, New Jersey: Allanheld, Osumn (1978), p. 4.
2 Cited in Wendell Berry, *The Unsettling of America, Culture and Agriculture*. San Francisco: Sierra Club Books (1977), p. 96. (Source: James E. Bostic, Jr., "Rural America, Where the Action Is," speech to the Farmers Home Administration State Conference, Raleigh, North Carolina, April 16, 1976.)
3 William Lockeritz, *et al.*, "A Comparison of the Production Economic Returns and Energy Intensiveness of Corn Belt Farms that Do and Do Not Use Inorganic Fertilizers and Pesticides," Center for the Biology of Natural Systems, St. Louis (1975).
For a more recent report, see Robert Klepper, *et al.*, "Economic Performance and Energy Intensiveness of Organic and Commercial Farms in the Corn Belt: A Preliminary Comparison," *American Journal of Agricultural Economics* (February, 1977), pp. 1-12.
4 Hall, D. C., R. B. Norgard and P. K. True, "The Performance of Independent Pest Management Consultants," *California Agriculture*, 29:10, pp. 12-14 (1975).
5 L. Watson, M. Galthouse, E. Dorothy, *Failing the People*, Agriculture Research Accountability Project, Ithaca, N.Y. (1972).
6 Perelman, *op cit.*, p. 39.
7 See Armand C. Lepage, "Soil Erosion and Agricultural Non-Point Source Pollution: Two Problems, One Solution," Office of Federal Activities, U.S. Environmental Protection Agency, Washington, DC (November, 1977).
8 Perelman, *op. cit.*, p. 40.

A major proportion of the people in the U.S. live in cities and thus it is here where significant changes can and must take place. In this piece David Morris outlines some of the components of those changes. His perspective is a valuable one, for as Director of the Institute for Local Self-Reliance (67) in Washington, D.C., he has been involved in some of the best work in community economics research, neighborhood-based recycling, urban agriculture, solar cell development and small business. (They once ran a successful sprout business in their basement). The work of ILSR can be followed in Self-Reliance *(54), while much of David's early thinking on the viability of urban communities is in* Neighborhood Power *(25), co-authored with Karl Hess.*

INDEPENDENT CITIES: CHANGING THE NATURE OF ECONOMIC DEVELOPMENT

By David Morris

Society is shaped by the way it uses scientific knowledge. America has used its human creativity —the source of science—to fashion a society where goods are produced far from where they are consumed, where it is more economical to replace people with machines than to create worthwhile work for them, and where a handful of large institutions make the rules, dispense the wealth, and sell the goods and services.

The appropriate technology movement attempts to realign priorities; it is a profoundly ethical movement. Instead of permitting science to be motivated purely by private greed or military defense, it assumes a vision of a new society. Its central principles are:

1. Society should minimize environmental costs.

2. Systems should be designed on a human scale.

3. Equity should be a serious goal in designing any system.

4. Systems should be democratic.

In sum, this vision of society stresses local self-reliance. It encourages the formation of closed systems, where production and consumption are closely linked. In Fritz Schumacher's words, it is a society where local production comes from local resources to meet local needs.

THE POTENTIAL FOR LOCAL SELF-RELIANCE

Local self-reliance is rapidly becoming not only technically feasible, but commercially attractive. In America's first 200 years, the economy evolved from a reliance on wind, animals and wood for energy and shelter to one based on iron, steel and fossil fuels. One effect of this transition was an increase in the size and scale of systems.

When steel, for instance, replaced wood as a major structural material in the last century, it required a much greater capital investment in factories, not simply for technological reasons, but also for reasons of profit efficiency as envisioned by capitalist planners. Plastics, reinforced concrete and other materials, which are rapidly replacing steel, require less technological capital and have so far avoided the concentrations of ownership which have characterized steel. Even within the steel industry innovations are pointing to decentralization. No less an authority than *Fortune* magazine noted recently,

> These famous old economies of scale, which demanded gigantic equipment and blocked the entry of small would-be competitors, are greatly diminishing in importance. Small companies emboldened by steel's "new economies" are streaming into the industry, setting up regional, even local, plants, splintering the business into smaller pieces—and making money.[1]

According to John Blair, the new technological advances are neither labor-intensive nor capital-intensive. They are, in his terms, "knowledge intensive." As he writes:

> With plastics, fibreglass and high performance composites providing high strength and easily processed materials suitable for an infinite variety of applications; with energy provided by such simple and efficient devices as high energy batteries, fuel cells, turbine engines and rotary piston engines; with computers providing a means of instantaneously retrieving, sorting and aggregating vast bodies of information; and with other new electronic devices harnessing the flow of electrons for other uses, there appears to be aborning a second industrial revolution which, among its other features, contains within itself the seeds of destruction for concentrated industrial structures.[2]

The thrust of modern science is potentially decentralizing. Modern developments in enzyme and microbe engineering, in the manipulation of atomic and molecular structures, in our understanding of life's basic forces, now permit us to consciously design our materials and our production and distribution systems to meet our ethical criteria.

Natural resource shortages, environmental legislation and rising energy prices make it economically attractive to use our scientific knowledge to promote local self-reliance. In the late 20th century the raw material base of our society may once again be changing, this time from steel, iron and fossil fuels to sand, plant matter and solar energy. Most relevant to our own discussion is that these three material resources are widely distributed throughout the globe, and lend themselves to a new sense of localism.

Sand, or silicon dioxide, is 27% of the earth's crust. It is the only material which is produced in every one of our 50 states. Silicon is the basic material for our electronics industry. It is also the building block for solar cells, tiny devices that convert sunlight directly to electricity. The use of silicate compounds is expanding rapidly. One report from the National Academy of Sciences concludes: "The crystal structure of silica and silicate minerals, each with attendant physical properties, already provides a basis for future substitution for the scarce metals."[3] Low density glass can now be made stronger than most metals. Silicate fibers and silicon carbide materials have already entered the industrial and consumer markets.

Plant matter already provides a basic structural material, wood, and our paper products. It can, as well, provide pharmaceuticals and plastics. During the Depression, Henry Ford demonstrated the feasibility of making plastics from soybeans. Some researchers believe that the rapid expansion of plastics could threaten the central role steel plans within the next generation.

Natural resource scarcity has prompted scientists to look toward increasing the yields of plant products. Currently three acres of timberland is necessary to provide all the wood products required by an American. New growing techniques could reduce this to ½ acre for construction materials, and less than 1/10 acre for pulp, paper and industrial chemicals.

The average American currently uses one to two acres to support his or her diet. Yet with recently demonstrated intensive farming techniques, as little as 1/8 acre could meet all but a small portion of this diet. Interestingly enough, our urban areas, contrary to popular belief, are not very dense. Manhattan, which most people envision when cities come to mind, has a population of 100 people per acre, but in cities of over 100,000 the density is rarely above 7 people per acre.

As we reduce our fossil fuel-based agriculture, we will need to make better use of the organic wastes that contain vital fertilizing nutrients. The human and organic wastes aggregated in our cities can be recycled into food production systems.

The 400% increase in the price of crude petroleum in 1973 made solar energy more attractive. Coal seams, oil wells, and uranium deposits are located unevenly around the globe, and lend themselves to private ownership and concentrated power. Sunlight, and wind, are everywhere and can be converted into useful energy by modern science. Recent studies in California found that a typical city of 10,000 could meet all but its industrial energy requirements through renewable resources. These estimates presume current consumption habits. Recycling, conservation and more efficient processes could dramatically reduce the resources needed to support our civilization, and thereby dramatically increase the potential for local self-reliance.

Our cities can become the base for the extraction of raw materials which can be processed economically in small factories serving primarily local market areas. The average size of a manufacturing plant in this country has not increased over the last two generations. It still employs fewer than 75 workers. One study of consumer goods manufacturing found that almost 70% of the industries for which a minimum efficient scale could be identified required capital assets of under one million dollars. If automobiles and petroleum products were excluded, 58% of the total goods needed by a population of one million could be locally produced in efficient small plants, while 16% of the needs of a market population of 200,000 could be met.[4] A factory could produce competitively priced, good quality photographic equipment and supplies for a market no larger than a big city neighborhood. Cambridge, Massachusetts, could support a hat or cap plant. Chicago could support a storage battery plant.

Even this rosy picture may be conservative. Several studies concluded that if we were to operate a factory only one-third the size of those cited above, the sales price of the finished product would rise only slightly. Schumacher's dream of local factories producing for local needs appears a genuine possibility.

At the same time, recycling technology is rapidly maturing. There are now paper mills that can use 100% recycled cullet. Modern steel mills can use 75-90% scrap metal. Our cities are the storehouse of wastes, and as recycling expands, our cities will be mined for their raw materials. Manufacturing plants tend to locate near their raw material supply, and this should prove no exception.

Recycling plants tend to be small. Factories which use recycled materials are less expensive to construct, can produce less and still operate economically, and can therefore serve smaller market areas. An aluminum plant using recycled aluminum costs only 1/3 that of one using virgin ores. Similar cost reductions can be seen in paper and glass facilities. Modern steel mills using scrap are often called mini-mills by the industry, or even neighborhood steel mills because of both their size and market area.

This analysis does not take into account the new materials advances touched on above. Plastics factories tend to be quite small, operating out of small shops with an average of 25 employees. Electric vehicles are far less complex to manufacture than existing internal combustion engine vehicles, and can also be produced in relatively small facilities.

The tooling required to produce a plastic bodied electric car, estimated in the late 1960s, was between $1.5 and $3 million, compared to the $500 million necessary to initiate production of conventional steel-bodied piston engine cars.

City: a settlement that consistently generates its economic growth from its own local economy.
—JANE JACOBS

We should not, however, jump to the conclusion that because the potential for decentralization exists, it will inevitably occur. Computers have not only made centralized information gathering possible, they have produced a global administrative network unparalleled in history. Solar cells can be placed on orbiting satellite systems, producing the most centralized form of electric generation in history. Resource recovery plants, metropolitan sewage treatment systems, large scale subway systems, all attest to the fact that science can be used, or misused.

Local self-reliance is technically feasible. But it is politically difficult. The rules of the game, written long ago, favor bigness and capital intensive systems. The investment tax credit and accelerated depreciation allowances make it easier to increase machinery size than to disperse into smaller, more labor-intensive operations. Federal research and development monies, representing 40% of all R&D in this country, go primarily to develop large-scale, capital-intensive systems of all kinds. Federal procurement procedures favor large companies. Depletion allowances favor the mining of virgin ores, rather than recycling. Freight rates tend to favor long distance transportation. Publicly supported interstate highway systems have long subsidized trucking. Declining block utility rates have promoted energy intensive manufacturing. In order to promote local self-reliance, in order to use modern science to fashion a society that meets our ethical criteria, we must change the rules of the game.

THE CITY, FREE TRADE AND SCIENCE

One could make an excellent philosophical argument for the value of urban areas. Cities have al-

160

ways been the centers of civilization, its storehouses of knowledge and innovation. They have been the driving force for creativity and progress. To the Greeks the city was a means to the self-realization of her citizens. Medieval cities were based on a strong sense of mutual aid, and a love of place. But we are interested here in another more pragmatic characteristic of cities: *power*.

Cities are unique institutions in this country. They are increasingly being given the legal and economic authority to change the rules of the game to encourage local self-reliance. They are the formal political unit of government closest to the people and therefore in the best position to redefine the function and duties of citizenship.

The issue of power cannot be taken lightly. To encourage local self-reliance and the ethical principles underlying appropriate technology means to change society radically. It means a 180-degree change in the doctrines which underpin much of our activity: free trade. It is ironic that the city, the birthplace of our modern economic system, should become the focal point in a struggle to restructure economics at the same time as it threatens to overwhelm our sense of community and independence.

The revival of cities in Europe in the Middle Ages was a difficult undertaking. Those who came to the city were often refugees from other parts of the tightly structured feudal system. The marketplace was both the city's reason for being and its defense. The Church and King supported the infant city in return for part of the surplus generated in the marketplace. For this revenue, the city exacted its own price. It rewrote the law to encourage a new type of organization, the corporation, and to support freedom of action and the enforcement of contractual agreements.

The protection and encouragement of business was made an integral part of the law. In Beauvais, France, in 1283, for example, the fine for attacking a citizen increased by more than 1000% if the citizen was in, or on the way to, the marketplace.

The early cities were relatively communal and egalitarian. Small merchants and craftsmen were often indistinguishable. But the dynamics of trade and private profit made themselves felt. Within the town, the corporation began to take on a life of its own.

> The medieval ideal of a town as the common possession of the inhabitants, descendants of those who had united upon their solemn oaths, was giving way. The "corporation" came to mean the town officials, the wealthiest and most powerful of the inhabitants, who treated the town's property as their own, selling it and pocketing the proceeds.[5]

The acquisitive appetite of this new creature soon spilled over the boundaries of the city. As trade became commerce, corporations needed a higher authority to enforce contracts and oversee business development.

> In order to accommodate the rationalization of production, fields had to be enclosed, guild privileges overridden in charters to companies of entrepreneurs and exporters, and local laws rewritten; the task required a powerful, centralized authority...[6]

The nation state fit these requirements. Individual cities, even leagues of cities, found that they could not cope with the exigencies of expanded commerce. Military defense and the need to generate sufficient capital to participate in expanded commerce and nascent industrialization dictated that the nation state was the most appropriate economic, legal and political form for the next historical period.

Nations became the protectors of business, and fought with other nations for access to markets.

> With the advent of manufacturing, various nations entered into a competitive relationship. The struggle for trade was fought by way of protective duties and prohibitions, whereas earlier the nations, insofar as they were connected at all, had carried on an inoffensive exchange with each other. Trade from now on had a political significance.[7]

In the United States the momentum of manufacturing and trade changed our very organization of government. The Articles of Confederation were overturned after only a brief period following the Revolution when the states began to use their powers to fragment the national marketplace. The new Constitution expressly forbade any interference in the flow of interstate commerce.

Throughout this historical process, science was the key factor which made expanded production and markets possible. Advances in shipbuilding and in navigation helped to expand both the quantity of goods to be traded and the area over which they could be safely transported.

With the advent of manufacturing, the budding discipline of engineering arose. Technical innovation followed the marketplace. In the mid-20th century, Jacob Schmookler looked back at technological innovation and, after many years studying its process, simply concluded, "most new industrial

technologies are found because they are sought."
He who pays the piper calls the tune, and for the
last 200 years the piper has been business. The tune
has been increased productive capacity, expanded
market areas and profits.

When the weavers in England fought against the
imposition of a factory system, the factory owners
turned to science and engineering. One sympathetic
contemporary wrote:

> The necessity of enlarging the spinning frame has
> recently given an extraordinary stimulus to mechani-
> cal science. . . . In doubling the size of his mule, the
> owner is able to get rid of indifferent or restive spin-
> ners, and to become once more the master of his mill,
> which is no small advantage.[8]

The author concludes, "This invention confirms the
great doctrine already propounded, that when capital enlists science in her service, the refractory hand
of labor will always be taught docility."[9]

By the early 20th century the fragmentation of
work had become the focus of increased scientific
attention. The assembly line and industrial engineering were born. The work force saw what was
coming, but were powerless against the increasingly
powerful concerns, supported by a government
which increasingly believed its only role was to enhance commercial opportunities.

The International Molders Union cried out against
the human inefficiency of the basic process underlying the new "scientific management" movement.

> The gathering up of all this scattered knowledge,
> systematizing it and concentrating it in the hands of
> the employer, and then doling it out again only in
> the form of minute instructions, giving to each
> worker only the knowledge for the performance of a
> particular relatively minute task. This process, it is
> evident, separates skill and knowledge, even in their
> narrow relationship. When it is completed, the
> worker is no longer a craftsman in any sense, but is
> an animated tool of management.[10]

Frederick Winslow Taylor, the father of the new
discipline, agreed. In 1919 he observed:

> Without the element of stupidity the method would
> never work. . . . A pig iron handler shall be so stupid
> and phlegmatic that he more resembles the ox than
> any other type. . . . The man who is mentally alert is
> entirely unsuited to what would, for him, be the
> grinding monotony of work of this character.[11]

THE LOGICAL CONCLUSION

By the middle of the 20th century the dynamic of
modern commerce and modern technology had
borne their logical offspring: enormously increased

production, significantly decreased worker satisfaction and wide environmental impacts.

A nation of producers had been transformed into
a nation of consumers. Work had become a way to
earn enough to consume (or for millions, to survive). It was no longer satisfying in and of itself.
The increased educational levels only served to
underscore the unskilled nature of the workforce.
Compare the theoretical knowledge of the craftsman at the time of the Industrial Revolution with
the capabilities of the modern office clerk, the most
rapidly expanding part of our labor force.

> Even the ordinary millwright . . . was usually a fair
> arithmetician, knew something of geometry, level-
> ling, and mensuration, and in some cases possessed
> a very competent knowledge of practical mathe-
> matics. He could calculate the velocities, strength,
> and power of machines: could draw in plan and
> section. . . .[12]

A detailed manpower survey done in the 1960s
by the New York Department of Labor revealed that
almost two-thirds of all the jobs in that state involved such simple skills that they could be, and
were, learned in a few days, weeks, or at most
months of on-the-job training.

Modern urban dwellers, despite new communications technology which gave them unprecedented
access to information, had come to be divorced from
any direct involvement in the world around them.
As Lewis Mumford comments:

> For the denizens of this world are at home only in
> the ghost city of paper: they live in a world of
> "knowledge about" . . . and they drift further away
> from the healthy discipline of first-hand "acquain-
> tance with."[13]

During this period the growth of corporate power
continued unabated. The major function of government was believed to be the enhancement of commerce. By the 1890s a distinguished legal treatise
could unabashedly declare, "[the] regulation of
business activity was no longer to be deemed a
proper function of the law of corporate organization. The function of corporate law was to enable
businessmen to act, not to police their actions."[14]

By the middle of this century, this hands-off attitude led to unprecedented levels of concentration.
By 1970, of the 275,000 manufacturing corporations
in the United States, only 200 owned over 65% of
all assets. Fewer than 30 giants owned over 20%
of all cropland. Eight oil companies controlled 64%

162

of proven oil reserves, 44% of uranium reserves, 40% of coal under private lease, and 40% of copper deposits.[15]

By the middle of the 20th century, these corporate giants had spilled over national borders. As two journalists commented, "So powerful are these giant economic enterprises that of the one hundred largest 'money powers' (measured by GNP or gross sales) in the world today, thirty-six are no longer even countries, but American multinational corporations."[16]

"For business purposes," says the president of the IBM World Trade Corporation, "the boundaries that separate one nation from another are no more real than the equator. They are merely convenient demarcations of ethnic, linguistic and cultural entities. They do not define business requirements or consumer trends. Once management understands and accepts this world economy, its view of the marketplace—and its planning—necessarily expand. The world outside the home country is no longer viewed as [a] series of disconnected customers and prospects for its products, but as an extension of a single market."[17] Professor Howard Perlmutter of Wharton School predicts that by 1985 200 to 300 global corporations will control 805 of all productive assets of the non-Communist world.[18] Union Carbide, one of the giants, could boldly announce in TV spots in 1976, "Today something we do will touch your life."

By the early 1970s, a historical watershed was reached. For the first time in history, a majority of global trade was no longer taking place between national corporations, but between branches of the same corporation. No longer was France trading with the United States. One division of General Electric was trading with another division of General Electric.

> Under one deal, GE do Brazil will sell $3.5 million worth of refrigerators to the Middle East this year. The ISG (International Sales Group of GE) is also trying to sell $20 million worth of Brazilian-made locomotives to Mozambique. GE do Brazil (in Sao Paulo), which has exported locomotives to other African countries and Latin America, is seeking Brazilian government financing for the sale. If the deal falls through, GE will try to sell locomotives to Mozambique from the Erie [Pennsylvania] plant instead.[19]

Thus, Sao Paulo and Erie's economies depend on how much the Brazilian government is willing to give to General Electric.

Our cities, once the proud birthplace of business, have become company towns, dependent on branches of farflung corporate empires. For, along with concentrated power came its corollary: absentee ownership. In Washington, D.C., Safeway supermarkets, headquartered in California, control 60% of all supermarket sales. In Maine, 13 of the 16 largest manufacturing enterprises are owned by interests outside the state. It is increasingly true that "America shops at Sears."

The mother corporation shifts its pieces on the global chessboard, making decisions based on the needs of its stockholders, not of the localities in which it is based. Anaconda Copper Company increases its mining output in Butte, Montana, when its Chilean-based subsidiary is purchased by the Chilean government. Later, when that government is overthrown, Anaconda reduces its Butte workforce. Youngstown Sheet and Tube Company, in Ohio, is purchased by a steamship company, and several years later its rundown plant is closed, 5,000 people are given notice, and an entire town teeters on the edge of bankruptcy. These plant closings are rarely done with advance notice. In the business world, one never provides creditors and competitors with more information than is necessary.

Right after the first layoffs in Youngstown, Mayor Thomas P. Dalfonso of Monessen, Pennsylvania, a city of 15,000 south of Pittsburgh, talked about the uncertainty: "You wake up and you're tight as a spring until you see what the headline is in the paper each morning." One day International Nickel Company laid off 3500 workers in Sudbury, Canada, to increase investments in Indonesia and Guatemala. Elmer Sophia, a member of the Ontario legislature, angrily commented, "[The community] willingly suffered the pollution of air, stream and lake, and even as INCO razed the countryside with its sulphurous fumigations the desecration was accepted, if grudgingly." He concludes, "The thrust to maximize its profits harbors no conscience. . . . Like all multinationals, INCO's business is business and gratitude is an unwelcome stranger to the board room."[20]

These branches are not always closed because they are losing money. They may be closed because their investment could earn a greater return elsewhere. Joseph Danzansky, formerly president of Giant Foods, and now Chairman of the Board, candidly conceded, "But let's face it. Many stores are closed not because they operate at a loss, but because they are marginal and the capital can be more advantageously invested elsewhere."[21]

The very goal of these businesses, as with any business, is the dependence of the city and its inhabitants on their products. Corporations are successful only inasmuch as they can either maximize their share of the market, or increase their profits. In achieving the first goal they make the cities increasingly sterile, company towns. In achieving the second, they withdraw money from the local economy, to invest it elsewhere where the highest return beckons.

ACTIVIST CITIES

Localities respond in varying fashions to these threats on their sovereignty. Many continue to vie with each other to attract new industry, giving away millions of dollars in incentives. The elaborate mating dance of city and corporation is a demonstration of the unequal relationship between them. The very bigness of business is its attraction. The short-term benefits of a 3,000-employee plant for local political officials outweighs any long-term considerations.

As even one business journal describes it, tax incentives set the stage for a "rising spiral of government subsidies as companies play off city against city and state against state for the most advantageous terms."[22] According to Joseph Hoffman, New Jersey's Commissioner of Labor and Industry, "It's cutthroat, regrettably, but it's every state for itself."[23] An ad in one paper extolled Michigan's supposed tax advantages: "Think it over. If your plant is in Ohio, Pennsylvania . . . Michigan is just a short distance away. A move just might be a very profitable move."[24]

Some local leaders recognize the problem, but believe themselves powerless. Governor Hugh Carey, describing the plight of New York, said, "We in New York are in the position of a nation that is running a severe balance of payments deficit. But we are not a nation unto ourselves, so most of the tools that nations use under these circumstances are unavailable to us. We can't devalue our currency. We can't impose import bans. We can't ask the International Monetary Fund for a loan. . . ."[25]

Others turn to the federal government for assistance. There is no doubt that federal policy has discriminated against our urban centers, and that financial redress is required. The resources of cities have been undermined considerably by federal programs and politics. But increased federal aid breeds its own type of dependence, that of the grant economy. Federal aid to the nation's largest cities rose more than tenfold over the past eleven years and amounted to nine cents of every dollar of locally generated revenue in 1967 and 50 cents on every dollar in 1978. One urban expert warns:

> The misfortune of the concentration on getting handouts from senior levels of government is that it leads to the avoidance of the question of whether at some level, local economies do not have to be self-sustaining. Clearly, neither the nation nor the states can survive if all beneath them is in the red.[26]

Still, a growing number of localities are fighting back. Cities are beginning to view themselves as basic planning bodies. The concept of planning itself has matured, as the role of the city in development efforts sharpens.

We may well look back to the middle of this century as the turning point in the centuries-old process of fragmentation and concentrated power. In the last 20 years, while the alienation of our citizens from the workplace and the political process has never been greater, the foundations for reconstruction have been laid.

The first step was to gain an accurate picture of the nature of resource flows in our cities. Computers provided the capability. Sophisticated statistical methodology and input models provided the means. Federal grants, contingent on local planning, provided the catalyst for data gathering.

The first portraits were a bleak picture of local hemorrhaging; the flow of local resources beyond local boundaries.

- In D.C. one McDonald's food store has $750,000 in annual sales. It earns $50,000 in profits and exports from the community between $500,000 and $567,000.

- Zip Code 60622, an older ethnic neighborhood in Chicago, found that it had deposited $33 million in a local savings and loan association, but had received back only $120,000 in loans. The deposits were being loaned outside the neighborhood, and often, outside the city itself.

- The low income neighborhood of Shaw Cardozo, in Washington, D.C., found that it was paying out more in fees and taxes to local government than it was getting back in welfare and services. Even the poor were subsidizing government.

These analytical tools have been helpful as well. Oakland and Hartford have developed skill inventories of the unemployed. Washington, D.C., found out that 20 percent of its space was in rooftops. Cities are using infrared cameras to identify heat loss in buildings for effective conservation programs. The composition of solid waste streams is being analyzed, as is the nutrient and metal characteristics of sewage wastes. The dynamics of winds and the intensity of sunlight are being explored. The air is being monitored for pollutants, and the soil is being tested. The mapping of our cities for self-reliance is well underway.

As the jigsaw puzzle of municipal resources is painstakingly pieced together, our understanding of the role of government has matured. In the 1960s, the consumer movement convinced the courts that the community could protect its residents from faulty and harmful products. Warnings appeared on soft drink containers and cigarettes. Match books were required to put their striking surfaces on the back. Millions of defective cars were recalled. The environmental movement forced us to go one step further to recognize that we needed to protect future generations from the consequences of our present actions. The social costs of private business have begun to be integrated into corporate balance sheets.

Federally required environmemtal impact statements have been copied by state and local governments. Tampa, for example, requires an impact statement on any new construction greater than $500,000. California requires that impact statements for development include an analysis of the feasibility of energy conservation and renewable energy resources. Madison has proposed legislation to require an analysis of the impact of rent increases on the tenants of existing buildings. A court upheld the need for an environmental impact statement to take into account the sociological impacts of density in the design for a new in-town development in Minneapolis. Private profit now has to be balanced against the social welfare. Moral considerations are entering the marketplace.

In the 1960s the federal government recognized the role that non-profit organizations can play in economic development ventures. Community development corporations and neighborhood development corporations were given legal authority to operate businesses which make a profit. There are now neighborhood governments which have subsidiaries which are profit-making businesses.

The boundary line between profit-making and non-profits has become increasingly indistinct.

Finally, a new generation has come of age and, as with any generation, it comes to power first on the local level. Although by no means homogeneous, the value system of this age group stems from a specific historical perspective. This group grew up during a period when 60 percent of the world became free of feudal lords and colonial masters (although not, unfortunately, of national despots). It was the period of civil rights, environmental concern, the Vietnam War, the women's movement, and an enormous focus on individual rights.

These developments set the context for the rise of activist cities. Two other developments gave these cities the potential for action: rapidly expanded budgets and enlarged political authority.

Without power and independence, a town may contain good subjects, but it can contain no active citizens.
—ALEXIS DE TOCQUEVILLE

Our cities need not be bankrupt. The balance of power has shifted from the national government to our localities. Between 1946 and 1974 the federal government increased its domestic spending by 700 percent, but state and local governments raised their own expenditures, exclusive of federal aid, by more than 1500 percent. In 1970 cities and counties employed almost three times the number of civilians as did the federal government.

In 1963 the federal government purchased $39 billion worth of goods and services. State and local governments purchased 25 billion, or 39 percent of the total. A decade later the figures were almost reversed. The federal purchases had increased to $54 billion, state and local purchases had expanded by 300 percent to $76 billion, or 58 percent of the total. The city of Washington, D.C., for example, with a population of 700,000, purchased about a quarter of a billion dollars in goods and services in 1977.

Direct aid from the federal government increased significantly, from about $7 billion in 1969, or 15 percent of the total expenditures by state and local governments to $80 billion in 1977, or about

28 percent of the total.

As federal largesse increased, the conditions under which it could be dispensed decreased. In the 1960s hundreds of conditional aid programs mushroomed. Cities developed proposals for federal aid. In the 1970s block grant programs were established. Money was allocated based on a complex formula taking into account local poverty levels, and tax rates, and population. The money needed to be spent on the low and moderate income population. Beyond that, cities were increasingly free to use the funds as they saw fit.

THE COURTS AND LOCAL SELF-RELIANCE

None of these historical developments would mean much, however, unless cities were being given the authority to change the rules of the game—to encourage local self-reliance at the expense of the free flow of goods and services across political boundaries. If appropriate technology means local self-reliance, a reliance on small business located near the consumer markets, and if our current tax laws, federal procurement policies, regulatory commissions and criminal codes discriminate against small businesses and local markets in favor of big production units and long loop distribution systems, then

when localities try to equalize costs and benefits they will be disrupting traditional trade patterns.

Interestingly, the courts have apparently been willing to extend to cities a vital role in local planning. There appear to be several justifications.

First, an increasing number of cities have gained home rule status under revised state constitutions. In states where cities gain their authority only from the state legislature, they are usually prohibited from doing anything unless it is specifically authorized by the state legislature. Home rule provisions reverse this procedure. Cities can do anything not expressly forbidden by, or in opposition to, state laws.

Second, the courts have supported cities in their contention that modern conditions require greater planning authority. As one Minnesota high court explained, "the need for local power grows with the complexity of modern life and the population."[27] A Court of Appeals noted, "In short, the police power, as such is not confined within the narrow circumspection of precedents, resting upon past conditions which do not cover and control present day conditions . . . that is to say, as a commonwealth develops politically, economically and socially, the police power likewise develops within reason to meet the changed and changing conditions."[28] Still

166

another court noted "economic and industrial conditions are not stable. Times change. Many municipal activities, the propriety of which is not now questioned, were at one time thought, and rightly enough so, of a private character."[29]

Third, a new environmental awareness has infected court doctrine. Chief Judge Brown of the 5th Circuit Court of Appeals frankly admitted:

> We hold that nothing in the statutory structure compels the Secretary to close his eyes to all that others see or think they see. The establishment was entitled, if not required, to consider ecological factors, and, being persuaded by them, to deny that which might have been granted routinely five, ten or fifteen years ago, before man's explosive increase made all, including Congress, aware of civilization's potential destruction from breathing its own polluted air, and drinking its own infected water, and the immeasurable loss from a silent spring like disturbance of nature's economy.[30]

In the late 1960s, Milwaukee calculated the impact of receiving 100 new families. It would require the availability of 100 dwelling units, 2.2 grade school classrooms, 1.64 high school classrooms, 4 teachers ⅔ of a fireman, 10,000 gallons of water and 140 new automobiles. In November 1972, Boca Raton, Florida, became the first city to impose a legal limit on its population, decreeing that no more than 100,000 persons shall be entitled to settle in the city. In 1976, the Supreme Court upheld the right of Petaluma, California, to limit future population growth. In the future, a person might own a plot of land within Petaluma city limits, zoned residential, and have a buyer from outside the city who wants to build a house on it, and the city can legally refuse permission to build based on environmental considerations.

In 1972, Palo Alto, California, rezoned almost 5000 acres within the city as open space that could be used primarily for agricultural purposes. The concept of private ownership of land has changed. Two legal scholars comment:

> If one were to pinpoint any single predominant cause of the quiet revolution, it is a subtle but significant change in our very concept of the term "land," a concept that underlies our whole philosophy of land use regulation. . . . Basically, we are drawing away from the 19th century idea that land's only function is to enable its owner to make money.[31]

Increasingly, the courts are entertaining not only the question, "Will this use reduce the value of surrounding land?", but also, "Will this make the best use of our land resources?"

The belief that the community can intervene in the freedom to use private landholdings in order to protect the larger welfare of its citizenry permeates other areas of local society as well. In Washington, D.C., a neighborhood parking ban on cars from outside the neighborhood, especially commuters, was upheld by the Supreme Court. In part of its opinion, the court declared, "The Constitution does not outlaw the social and environmental objectives, nor does it presume distinctions between residents and non-residents of a local neighborhood to be invidious."[32]

Bowie, Maryland, was upheld by the courts in imposing an ordinance to require all bottles sold within its boundaries to be returnable. The court concluded: "The putative local benefits of the Bowie ordinance clearly outweigh any burden which the ordinance might impose on interstate commerce. The benefit which the ordinance is designed to achieve is a substantially cleaner environment."[33]

The city of Detroit, and several cities in New Jersey, have been legally upheld in their authority to require public employees to live inside city limits. The cities argued that public tax money was being spent outside city limits, and that the city had an obligation to reduce leakage of public money and recycle as much as possible into the local economy. Chief Justice Weintraub, in a New Jersey decision, stated, "Government may well conclude that residence will supply a state or incentives for better performance in office or employment as well as advance the economy of the locality which yields the tax revenues."[34]

In a recent Supreme Court case where the State of Maryland discriminated against out-of-state processors of scrap metal, Justice Lewis Powell commented, "Nothing in the purposes animating the commerce clause forbids a state in the absence of Congressional action from participating in the market and exercising the right to favor its own citizens over others."[35]

A number of cities have given preference for locally produced goods and services. Although the courts have frowned on direct discrimination of out-of-city suppliers, Oakland's City Council is studying the possibility of favoring local businesses when there is no more than a one percent difference in the bid. Hartford, Connecticut, and Detroit are doing this as well. The increasingly sophisticated economic methodology which can point to the very beneficial impact on the local economy of such a provision may lead to a change in legal thinking.

Recent federal legislation has encouraged cities to develop a sense of local resources. In 1977, through the efforts of hundreds of neighborhoods around the country, the Community Reinvestment Act was passed. Although not a local measure, it gives support to the concept of local self-reliance. It was a response to the redlining activities of banks and savings and loan associations. Savings and loans were established in the Depression to provide neighborhood-based financial institutions which would loan money for home ownership within small geographic areas. By 1975, as we noted above, one neighborhood in Chicago found that it had received no first mortgages for the $33 million it had deposited in the local savings and loan.

The new federal legislation requires federally chartered financial institutions to divulge where they put their money. If these financial institutions desire federal approval for some action (e.g. to open up a branch), they must prove they have been responsive to community needs. Thus, capital, that most abstract and mobile of all commodities, has also been burdened with moral constraints. Morality once again has been imposed on the marketplace.

These same trends are happening in environmental regulation. The Environmental Protection Agency has mandated air quality standards. Those cities which do not meet those levels by 1980 will not be permitted to have any new development which adds to pollution unless they first reduce by an equal amount their pollution levels in another area. The health of citizens is being balanced against the need for growth.

The conception of the city as a cooperative for the improvement and development of the capabilities and lives of its inhabitants is at odds with the doctrine of laissez faire and a national capitalism that has turned local citizens into consumers and so many free floating factors of production to be assembled and disassembled by the forces of the national market. The older conception of the walled city as a shared common enterprise has been weakened by the breaching of its walls and its transformation into an open economy.
—NORTON E. LONG

ACTIVIST NEIGHBORHOODS

At the same time as states and federal courts have given cities increased legal tools, the cities and local courts have given neighborhoods increased recognition.

Neighborhoods have been the forgotten child of cities. As private corporate power centralized outside the city, public corporations centralized power inside the city. During the last 100 years the trend has been for city hall to acquire more and more power. Urban reformers in the early part of this century criticized the neighborhood-based ward system and fought a successful battle to replace local representatives with at-large, or city-wide city councilors. It also promoted the city manager form of government, giving increased authority to a professional, unelected, manager of city operations.

But the 1960s, with urban riots and federal antipoverty programs, changed the role of neighborhoods. Those neighborhoods which had survived the onslaught of urban renewal and freeway programs in the 1950s and 1960s banded together to demand a positive role in planning efforts in the 1970s. They have been increasingly successful.

In Dayton, Ohio, neighborhood priority boards dispense community development money. In Atlanta, the city gives money to neighborhood planning units to hire technical assistance. In Washington, D.C., advisory neighborhood commissions are formally elected, and recognized by the city. In 1978, a District Court in Washington, D.C., required that all city agencies give proper advance notice to the advisory commission on all relevant matters before it, and that it give "great weight" to the neighborhood's recommendations. This means city agencies, such as the zoning commission and liquor licensing board, can overrule the neighborhood, but must give adequate justification for doing so.

The giant Housing and Urban Development Administration, under continuing pressure from neighborhood organizations, has switched its focus from urban renewal to neighborhood revitalization. The newly formed National Association of Neighborhoods represents growing numbers of neighborhoods which have changed their self-image from that of civic associations to that of neighborhood governments. A National Commission on Neighborhoods has been established by the White House.

Neighborhoods, through the federal recognition of the role non-profits can play in economic development, are combining economics and governance.

In St. Louis, a neighborhood owns a radio station. In Baltimore, a neighborhood is a housing developer. In Washington, D.C., a neighborhood owns and operates a garbage collection service. In Eugene, Oregon, a neighborhood does land banking.

So far, we have focused on the provision of goods; for manufacturing and raw material extraction, the city or county has the necessary market size and co-ordinating ability. But the provision of services has a different dynamic. The economies of scale of service delivery systems make the neighborhood an increasingly attractive base of operations.

Almost all of the services presently provided by larger institutions can be delegated to the neighborhood level. An important factor in the development and delegation of service delivery systems is the distinction between prevention of problems and treatment of problems. In prevention, there is a greater role for smaller systems, less capital expense, a greater role for paraprofessionals, and increased citizen involvement. Prevention systems, whether they be in health, criminal justice or other areas, tend to be decentralizing.

Treatment systems, on the other hand, tend to need capital intensive operations, highly trained professionals, and centralized facilities. The best place to prevent and maintain health is in the neighborhoods, where public schools and small clinics can educate people, diagnose problems, and talk to people about diet, living habits and stress. Only if we prefer to treat diseases *after* they occur are emergency rooms, expensive medical technology and highly trained personnel required. Neighborhood courts, family dispute arbitration, and people on the streets are the way to prevent crime. Only if we want to deal with crime after the fact are computerized police units, downtown court houses and faraway jails required.

Neighborhoods can operate medical clinics. The recent proliferation of health maintenance organizations indicates that such clinics may be a way of life in our future.

Neighborhoods can operate basic financial institutions. Geographically based credit unions were formally approved in the 1960s. For the first time, people could become members of a credit union merely by living in the same area. In 1977, the federal government allowed credit unions to issue checking accounts which gave interest. In 1978, credit unions were given the ability to give long-term home mortgages. Thus credit unions are being given increasing

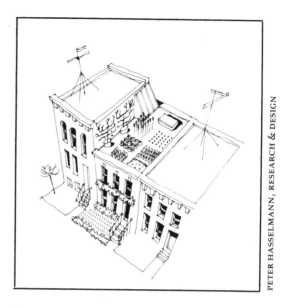

PETER HASSELMANN, RESEARCH & DESIGN

authority in areas previously restricted to banks and savings and loans. The governing principle is still one member, one vote, and their orientation is still towards their membership.

Neighborhoods can also provide the inspection services for a city. Housing inspection, restaurant inspection, and other responsibilities can be readily delegated to trained neighborhood residents. Neighborhood team policing is a concept which has been tried in a number of cities with success. Police are once again walking the streets, assigned to specific neighborhoods. In Dayton, Ohio, neighborhoods have, with police training and support, put their own teams of "police" on the streets and have been responsible for a majority of the arrests for minor crimes in the neighborhoods.

The neighborhood can supply a significant amount of the services that the city now allocates. It could provide garbage pick-up, it could be the prime contractor to subcontract for street paving and sidewalk repair. It could remove snow, plant trees and monitor noise and air pollution.

A SYNTHESIS: A NEW WAY OF THINKING

Over the past 20 years a quiet, and not so quiet, revolution has been taking place in this country. Modern science has given us the potential to decentralize. Cities have gained the legal tools and economic strength to grapple with modern problems. The resource base of cities is being mapped, and the role of the community in defining the rules of the marketplace is becoming central.

The citizens of this country have expressed a strong opinion that the future should look very different from the past. In May 1977, a Harris poll revealed that almost two-thirds of Americans believe we would be better off if we emphasized "learning to appreciate human values more than material values," rather than "finding ways to create more jobs for producing more goods." By an equal majority Americans preferred to "break up big things and get back to a more humanized living," over developing "bigger and more efficient ways of doing things."

The tools are there. Citizen support is there. We can now speak of independent cities. By linking up the creativity of our citizens with the legal and economic power of our municipalities, the concept of local self-reliance can become a reality. But it will take a new way of thinking.

In 1977 Mayor Abraham Beame of New York City went on television to criticize the Environmental Protection Agency for attempting to "strangle" New York's economy by banning cars in downtown Manhattan. The mayor, however, missed EPA's intention and overlooked an excellent opportunity for local self-reliance.

EPA was banning not cars, but the internal combustion engine. It was doing so because air pollution levels in New York were killing its citizenry. EPA did *not* ban electric vehicles. In 1976, in hearings before Congress, manufacturers indicated that an electric car manufacturing plant could be economically established for under $3.5 million. It would operate economically if it produced between 7,000 and 12,000 vehicles per year.

New York has sufficient money from its Community Development Block Grant program to capitalize the venture. Its own agencies purchase thousands of vehicles per year. It has both an in-house engineering staff and staff on city universities who can begin to do the research and development tasks necessary for an infant industry.

Electric cars would reduce noise and air pollution. One study indicated that in Washington, D.C., air pollution adds about $300 per year to an average citizen's cleaning bills. This does not count the lost work hours due to headaches and other pollution-related diseases. Electric vehicles are lighter and produce less wear and tear on local highways, whose maintenance comes out of local budgets. They lend themselves uniquely to use by the physically handicapped. Thus, electric vehicles could

generate jobs, save dollars, improve the environment of the city, and the new industry could be generated by an efficient and integrated use of a city's resources.

Indeed, it is probable that New York City could have gone even further in planning for a new transportation industry. A study done by the Institute for Local Self-Reliance for Washington, D.C., found that, if all workers used electric vehicles for commuting, and if all parking garages and parking lots were covered with solar cells, those vehicles could gain 65 percent of year-round transportation-related energy from these solar cells, and almost 100 percent in the spring and summer months.

We can use New York for another example, although this one could as well apply to many other cities. In 1978, New York's wastes were threatening to overwhelm it. Sewage dumped in the ocean was creeping back onto its beaches. The landfills in the Bronx were overflowing, filled way beyond their planned capacity. At this writing the city is about to enter into a contract with a private company to handle all its wastes. The city would transport all the garbage generated in the Bronx to a central facility. The private firm would then burn most of the garbage, and would sell the energy back to the city. The city would pay to have the company convert the garbage to fuel and would pay again to get the fuel back as energy.

Instead, the city could look to its wastes as the raw materials for a new economy. It could, for example, separate out newspapers from the waste stream at the neighborhood level as is being done in several cities around the country and use these to develop insulation manufacturing plants. A city of 75,000, or certainly an area the size of the Bronx, with over two million population, can generate sufficient newspapers to run an economical cellulose factory. The city could give the neighborhoods a fee for persuading their citizens to separate out newspapers, thus generating revenue for future community projects. An insulation plant would need about $400,000 in capital investment. The city could use both its community development money for capitalization and its development corporations for financing. It could purchase the insulation for use in its own energy conservation projects.

The central issue is clear. Does the city want to act as an economic planning authority or not? We do not presume that the city would operate the plants. This is better left to the private sector. But the city would establish the rules under which such a plant would be operated. In return for its purchasing power, financing and control over the raw material supply, a city can require that certain social goals be met.

Other evidence of a new way of thinking relates to our concept of the educational system. Currently students are defined as consumers. They consume knowledge for more than a dozen years, gain a credential, and often move away from the community to become productive members of society. The society defines itself as a commercial entity, and students' needs are seen accordingly. Students need credentials, self-discipline and a basic package of marketable skills to fit into the mainstream.

Unfortunately, students are not taught survival skills. Every school curriculum should have, as a matter of course, instruction in basic wiring and plumbing, auto maintenance, sex education, basic health, simple legal processes and construction skills. An enormous amount of human talent is being wasted: 90 percent of our medical visits are for maladies which will disappear within a few days regardless of a doctor's intervention, and which could be easily treated by paraprofessionals or a bowl of chicken noodle soup; 90 percent of plumbing calls are for plugged toilets and leaky faucets; and 90 percent of electrical calls are for frayed wiring or blown fuses. Many problems can be handled without costly professional help if these skills were part of our educational curricula.

In Lehigh College, in Pennsylvania, a philosophy professor, teaching a course on self-reliance, had the students rehabilitate a house and sell it at the end of the semester. When asked, the professor said he wasn't sure if he were training philosopher-carpenters or carpenter-philosophers. But the concept of self-reliance, of linking theory and practice, of defining students as producers, had become clear to the students.

There is no reason that the academic curriculum could not be re-oriented toward local self-reliance. Trigonometry classes could use the sizing and orientation of solar collectors as their work project. Economics classes could evaluate the multiplier impact of local expenditures for different goods and services. Civics classes could teach about neighborhood governments and city charters. Biology classes could produce food and fiber. Science classes could redesign existing facilities for energy conservation.

When viewed from this perspective, our educational resources look different. The student body, numbering thousands of people in our cities, becomes the workforce. The faculty, with their years of experience and skills, become the supervisory force and the technical consultants. The dozens of laboratories and machine tool shops, the empty buildings after school hours, become the capital base. Students, in Buckminster Fuller's elegant phrase, could start to "learn a living."

Another step toward a new way of thinking is to accept that any decision must require an evaluation of its costs and benefits, and that democracy demands that citizens be fully aware of these costs and participate in the decision making process. In an era of limited resources, we must choose.

For example, recent studies indicate that half of the 450,000 Americans hospitalized as a result of heart attacks each year could be discharged from the hospital a week earlier than is usually the case. This would save money, and the stress attached with prolonged hospital visits. Yet the statisticians tell us that there would be a slight increase in deaths. For a city like Washington, D.C., it may mean the saving of about $1.5 million a year, at the cost of having one to three people die. Will the community accept this tradeoff?

To use another example: cities currently get a significant portion of their money from services. If property values are rising, assessments rise, and tax revenues increase. On the other side of the budget, large low-income families put a burden on city social services for increased welfare, medical expenses, education and crime prevention. Thus, if a large low-income family is replaced by a small professional family, the city gains twice. Property tax values, and therefore revenues, increase: service demands go down and expenditures decrease. From the city's perspective, the displacement of low-income neighborhoods, and the rising price of local real estate is a benefit.

From the perspective of a neighborhood, especially one which is stable and relatively low-income, such a development means the complete disruption of the community. How do we balance the need of the city to increase its revenues to provide adequate services with the needs of its neighborhoods to have an assured future?

One of the basic underpinnings for democracy is information. Cities can use modern technology to permit citizen participation. Unfortunately, two-way cable television only scratches the surface, by showing people city council meetings and having them press a button to vote on various issues. Information access can, and should, go far beyond this. The city has the tools to install terminals in people's homes and provide the information that will make effective planning at the local level possible.

Currently people could install a computer terminal and pay for it with the money saved from listing their names in the white pages and yellow pages of the telephone books. Our public service commissions could order phone companies to install such terminals. This would mean not only that people would no longer have to dial information for phone numbers, but that they could find out the closest florist to their home, the doctor who had hours on Saturday, the optometrist who had the lowest prices, the Better Business Bureau rating of a product, or the ingredients of new food products.

The city could link its computer tapes on home ownership and land use to citizen terminals. Thus, a neighborhood could begin to pinpoint when it begins to go into decline, and when it was prone to real estate speculation. People who find their landlords recalcitrant in making repairs could find out if the owner is also remiss in other buildings in the city.

We are moving toward an information-based society. There are some recent studies that indicate the majority of our workforce is currently involved in information-related activities, whether they be teachers, file clerks, secretaries, lawyers, librarians, doctors or the like. As accessible information becomes available to average citizens, and as schools begin to teach people the basic skills necessary to be paraprofessionals in various fields, the reliance on professional organizations will be reduced.

In the final analysis, of course, we must link up our resource base with our labor. In Hartford, Connecticut, the 1974-75 recession generated two related problems. The city needed revenue to pay for the increased welfare and other costs that the recession generated. It needed to raise property taxes at a time that increased unemployment meant that more people would be unable to pay their taxes. The city came up with a unique solution. It developed a list of the jobs necessary to run the city, and put an hourly wage on each job. Residents could choose to work off their property tax obligation.

Such a system, if expanded, could produce interesting development. Conservative politicians repeat the refrain that we now work two days a week for the government. Currently government is divorced from the community. But if we were to work directly for the community for one day a week, doing the jobs that needed to be done, the implications for citizenship might be extraordinary.

The community could analyze the work that needs to be done to maintain the city. The work could then be broken down into skill categories and the degree of training required. A skill survey of those in the city would be developed, and a resource map completed. The city would then permit its residents to work off their obligation by doing these jobs. Once again, in the Greek sense, the city could become a means for the self-realization of her citizens. Residents could gain a much more accurate understanding of what it takes to run the city. They could eliminate that work that is unnecessary, develop more efficient processes for that work that is required, and be a part of the total effort of planning.

One sober word of caution. The transition to local self-reliance will not be an easy one, for it undermines the enormous economic and political power of our giant corporations. The small towns of Elbow Lake, Minnesota, Hankinson, North Dakota, and Aurora, South Dakota, set up a municipal distribution system for electricity when the retail franchise of the Otter Tail Power Company expired. Otter Tail refused to sell energy to the new system at wholesale prices and refused to permit its wires to be used to wheel in power from other suppliers. Although the Supreme Court upheld the towns in their suit, it was a 4-3 decision, and the Chief Justice dissented, saying in part:

> As a retailer of power, Otter Tail asserted a legitimate business interest in keeping its lines free for its own power sales and in refusing to lend a hand in its own demise by wheeling cheaper power from the Bureau of Reclamation to municipal consumers which might otherwise purchase power at retail from Otter Tail itself.[36]

In the middle 1970s, when the city of Bowie, Maryland, initiated its returnable bottle bill, several bottle companies threatened to boycott the city. Dependence breeds a certain arrogance. City activists must recognize the political realities of independence.

CONCLUSION

Two urban observers, at the beginning of the 1970s, asked, "The question arises: could we, or should we, try to reconstitute the older style of spatially defined communities?"[37] The answer is yes. We have the technical capability to do this. Our new awareness of environmental costs urges us in that direction. Our need for a renewed sense of citizenship and community demands such a development. The city as a nation is a metaphor for a new society —one which is built on the creativity of its citizens, one which transforms a nation of consumers back into a nation of producers, one which uses political authority to encourage the application of science to develop humanly scaled systems.

In 1950 or even 1960, we might have embraced the theory and philosophy of local self-reliance, but we did not yet have the gathering of historical and technological forces that made it both practical and attractive. In 1980 we do.

Yet we should not underestimate the forces moving in the opposite direction. The struggle in the 1980s will be whether we will have a global village or a globe of villages. Will we continue on the road to centralization under the doctrine of a free trade unencumbered by social costs? Or will we move in another direction, toward a humanly scaled society where cities once again become the driving force for creativity and progress? ∎

1 Cited in David Morris and Karl Hess, *Neighborhood Power: The New Localism*, Boston: Beacon Press (1975), p. 120.
2 John Blair, *Economic Concentration: Structure, Behavior and Public Policy*, New York: Harcourt, Brace, Jovanovich (1972), p. 151.
3 Edwin W. Tooker, *National Materials Policy*, National Academy of Sciences (1974).
4 Barry Stein and Mark Hodax, *Competitive Scale in Manufacturing: The Case of Consumer Goods*, Cambridge, Center for Community Economic Development (1976).
5 Michael E. Tigar and Madeleine R. Levy, *Law and the Rise of Capitalism*, New York: Monthly Review Press (1977), p. 193.
6 *Ibid*.
7 Maurice Dobbs, *Studies in the Development of Capitalism*, as cited in Tigar and Levy, *op. cit.*, p. 185.
8 Andrew Ure, *The Philosophy of Manufactures or an Exposition of the Scientific, Moral and Commercial Economy of the Factory System of Great Britain* (London, 1935), p. 366, as cited in David Dickson, *The Politics of Alternative Technology*, New York: Universe (1975), p. 80.
9 *Ibid*.
10 Cited in Robert F. Hoxie, *Scientific Management and Labor*, New York: Kelley (1918), p. 94.
12 David S. Landes, *The Unbound Prometheus: Technological Change and Industrial Development in Western Europe from 1750 to the Present*, New York: Cambridge University Press (1969), p. 63.
13 Lewis Mumford, *Culture of Cities*, New York: Harcourt, Brace, Jovanovich (1970), p. 257.
14 James Willard Hurst, *The Legitimacy of the Business Corporation in the Law of the United States, 1780-1970*, Charlottesville: University Press of Virginia (1970), as cited in Kenneth Fox, *Better City Government: Innovations in American Urban Politics 1850-1937*, Philadelphia: Temple University Press (1977).
15 Jeremy Rifkin and Ted Howard, *Who Shall Play God?* (New York: Dell (1977), p. 105.
16 *Ibid*.
17 Richard J. Barnet and Ronald E. Muller, *Global Reach*, New York: Simon and Schuster (1974), p. 14-15.
18 *Ibid*.
19 *Business Week*
20 *Vancouver Sun*, October 27, 1977.
21 *Washington Post*
22 *Business Week*, June 21, 1976.
23 *Ibid*.
24 Edward Kelly, *Industrial Exodus*, Ohio Public Interest Campaign, Conference on Alternative State and Local Public Policy (1977), p. 18.
25 *New York Post*
26 Norton E. Long, *The Unwalled City: Reconstituting the Urban Community*, New York: Basic Books (1972), p. 69.
27 City of Saint Paul v. Dalfin, Minn.
28 89 Cal. Rpts at 905, Citing Miller v. Board of Public Works, 234 P. 2d 381 at 383.
29 Central Lumber Co. v. Waseca, 1922, 152 Minn. 201, 188 N.W. 275.
30 Zabel v. Tabb, 430 F. 2d 199, 200-201 (1970)
31 Fred Bosselman and David Callies, *The Quiet Revolution in Land Use Control*, Washington, D.C.: Council on Environmental Quality (1971), p. 314.
32 *Washington Star*, October 11, 1977.
33 Bowie Inn v. City of Bowie, 274 Md 230, 335 A. 2nd 679, note 10 at 6897.
34 Skolski v. Woodcock, 149 N.J. Super 340, 373 A. 2nd 1008, 1977.
35 Huges v. Alexandria Scrap Corporation, 426 US 794 at 810.
36 Otter Tail Power Company v. United States, 935 Ct. 1022, 1973.
37 Scott Greer and David Minar, "The Political Side of Urban Development and Redevelopment," *Annals of the American Academy of Political and Social Science*, Vol. 352, p. 67 (March, 1964).

Enough is known about the use of renewable sources of energy to begin to see how they can best be applied to provide more than just power. An important new option is community-scaled energy systems, which may prove to be particularly efficient and economically advantageous to locally manufacture and use. Lee Johnson is a person whose writing and consulting work with architecture/engineering firms, state agencies and utilities is based on hands-on experience with alternative energy systems in general—particularly wind machines and solar workshops. He is a long-time editor of Rain *(48).*

NEIGHBORHOOD ENERGY: DESIGNING FOR DEMOCRACY IN THE 1980s

By Lee Johnson

THE MOVEMENT toward energy self-help and decentralization is now about ten years old. We have come a long way from "alternative sources of energy," that once youthful, now odd-sounding rallying cry of the late '60s and early '70s that stemmed, in part, from such influences as TV's "Mr. Wizard" on budding junior scientists and engineers, the "do-your-own-thing" New Age homilies of a once rebellious but now usefully co-opted counterculture, and the desire to use applied technology more benignly to "make the world work for everyone" à la Buckminster Fuller. Nowadays, renewable, on-site solar energy is the "conventional wisdom," buttressed by the rediscovery and increasingly rapid application of eco-system survival principles, the energy-entropy laws and successful working examples to national energy policy. They are the only way to go. The truly "alternative" energy sources are actually nuclear and fossil power, in the sense that they should be used only as last resorts and with open recognition of their clear peril to our nation's security, our economic strength, our personal health and safety, world peace and the global environment.

Indeed, nuclear fission is now viewed as the world's most undemocratic form of energy. Yet more than that, it may simply be America's first inherently unconstitutional energy source. Although Aldous Huxley was one of the few early voices warning us of nuclear-powered totalitarianism, no less a nuclear energy advocate than Dr. Alvin Weinberg, director of Oak Ridge National Labs, agreed that we truly face a "Faustian bargain" with the devil. Both felt that only a rigidly controlled society could arrive at the perfection, amidst human error and malice, needed to run atomic power systems safely. Fortunately, the loss of our constitutional rights of free speech, press and assembly in the nuclear crucible, as noted by Hazel Henderson, Amory Lovins and others, need not occur. We've got plenty of other paths to walk.

For nowadays, a large and growing number of newly confident solar, wind, biomass and water power researchers, inventors, developers and businesses—once unwanted outcasts—are eagerly sought prophets and practitioners of more humane and life-enhancing energy technologies. And, after SUN Day, solar power has become too obviously popular for pro-nuclear bureaucrats in the U.S. Department of Energy to ignore by "studying it to death," or for any president to reduce its funding or continue shortchanging the sun's budget relative to

174

the budget for the nuclear cul-de-sac, as illustrated in Table 1.

FY '79	Budget ($ in millions)	DOE Employees
Nuclear	1,120.3	450 HQ 1,300 Field
Fossil Fuels	?	?
Solar Thermal	82.0	11 HQ 6 Field
Wind	40.7	4 HQ
OTEC	33.2	6 HQ
Biomass	26.9	3 HQ
Photovoltaic	76.1	8 HQ
Geothermal	126.7	55 HQ 15 Field

Table 1. Dept. of Energy Budgets and Staffing

Think about it. Have you ever before heard of nation-wide demonstrations by 20 to 30 million Americans in favor of the greatly increased use of a specific energy technology? Certainly this citizen action cannot be considered neo-Luddite, or anti-technological. All of these new facts, potentials and attitudes about solar energy have led to the formation of a strong and knowledgeable "Solar Coalition" of Congressional representatives, senators, their aides and staff members who support the rapid implementation of renewable energy with needed legislation, increased budgets, larger staffs and action directives. This important political shift will continue even more rapidly in the future.

Yet more than all of this has happened on our way to national energy sanity.

INTUITION, TRUST AND THE NETWORK

Early on we had only intuition—our "gut" feelings—to go on. It somehow just seemed "right" to turn on to the sun at the end of a turbulent decade marked by "getting your shit together," a rising environmental consciousness, and a "guns versus butter" debate won by a senseless foreign war. After all, if a succession of federal administrations could quash traditional liberties to involve us in a debilitating conflict, and lose our confidence with continuing lies about "light at the end of the tunnel," then our energy supplies might also be something the government could not be trusted to deal with. Again, it was up to a minority of concerned citizens to lead the nation to better ways.

At first, young science-oriented pioneers and old tinkerers labored with meager resources in a tiny national network of do-it-yourself experimenters communicating their successes and failures through personal visits and a few small circulation newsletters. Trust developed among former strangers who shared the same visions, similar simpler life-styles, personal values and "Yankee ingenuity" technical insights over long distances. By the time of the 1973 oil embargo, there existed a tightly knit group of experienced solar practitioners who were worried about national energy directions and ready to speak up a bit louder. Yet it seemed to be dangerous and foolhardy to do so.

TECHNOLOGICAL JEFFERSONIANISM

Like other numerically weak and unconfident minorities, the early alternative technologists took a paranoid stance and a purity of position in relation to the "establishment." If the energy thesis of the military-industrial-academic complex was high-tech rare earth minerals, elitist engineering expertise, megawatts of central-station nuclear plants and unresponsive utility executives, then the antithesis had to reflect funky garbage materials, garage workshop knowhow, kilowatts right in your own back yard, and the individual freedom of disconnecting from the hated electric grid. It was Jefferson's hardy and independent "yeoman in the woods" reborn in the heads and hands of energy activists nationwide.

At its most community oriented, the attitude was "Let's you and I get a small group of friends together, go live someplace away from this sick society and its evil system, and build our own energy system that *we* can control." Yet the technical actualities were still to be personal home energy systems even within such socially cooperating settlements. Partly this occurred because individual energy efforts were a new enough first step that had just been blessed by peer-group approval, because many working models existed across the land, and because information was more widely available on how to accomplish it. Small-scale systems were also easiest and safest to tackle first, and often not enough people of like mind and life-style lived in

the same place to do community energy systems. Initially lacking the power to affect even local energy decisions, individual options were the only feasible focus for development. They simply had more vital elements together. As a result, community energy lost out by default in the early attempts at collective self-reliance.

THE INTEGRAL URBAN NEIGHBORHOOD

Recently some conceptualizers such as Tyrone Cashman, formerly of the New Alchemy Institute-East (Cape Cod and Prince Edward Island) and now at California's Office of Appropriate Technology (OAT), have recognized that, given the successful model provided by the Farallones Institute's Integral Urban House in Berkeley, California, the next real step is not simply solar suburbia made of "American Dream" tract houses slightly mutated to tap the sun. Rather he suggests that a better direction to go is an entire urban neighborhood of homes that are energy-conserving and -gathering (with solar water and space heaters), food-producing (with back yard gardens, attached solar greenhouses, rabbit hutches, hen houses, fish ponds and bee hives), waste recycling (with compost toilets that turn human and kitchen wastes into fertilizer for the garden), water conserving (with waterless compost toilets, low water use showers and appliances), and materials reusing (with the homes themselves built of salvaged lumber, windows and other fixtures).

Yet a neighborhood of only such homes would be as "integral" as the sterile monoculture of commuter suburbia. Our task is not to create an isolated, walled fortress for groovy-hip counterculture devotees. A true integral neighborhood goes beyond single- and multi-family detached dwellings, like the Integral Urban House, or condominiums and apartments. A vibrant, life-enhancing real-world community contains commercial shops, restaurants, churches, taverns, stores, hospitals, professional offices, theaters and inns.

The earlier cul-de-sac of "residential only" personal energy independence should remind us to relate solar energy to the world we work in as well as that of homes and family.

COMMUNITY ENERGY: THE NEW SYNTHESIS

More than five years have passed since such technological individualism was the given currency of opinion among the hardware and philosophy leaders in appropriate technology. Residential solar energy systems independent of the electric grid, or of other similar nearby solar energy systems, will certainly continue to be built, but it is now widely agreed that cooperative energy efforts are generally much more cost effective than isolated attempts at self-sufficiency. Community-scale energy sources seem more conducive to realizing the social and economic ideals of our democratic society. For the new solar yeomanry of the '60s, when extended to its energy-isolationist extreme, sometimes produces bad effects, especially when abetted by America's highly competitive ethic. There is a fine line between energy independence that means one is less of a burden to the rest of society, strengthening everyone's ability to survive and prosper, and the "I've got mine" attitudes of some earlier pioneers that weaken the trust of mutual goodwill our social fabric is tenuously built upon.

Others have realized that since electric utilities are granted a publicly regulated monopoly by and for the benefit of a state's citizens, "they" don't legally own the electric grid. Rather, the citizenry, through their public utility commission, indirectly control it and can mandate its use, by the electric companies, to include aiding and promoting all individual and cooperative efforts to feed power back into the grid to resell. Essentially, the potential benefits to all citizens of an electric distribution network constructed at their sufferance, should not be ignored and vilified but rather put to use. The grid is ours, not theirs!

Yet if we are to design for local democracy in America's energy future, we must understand not only the straightforward technical means making such efforts possible and the economic reasons making them sensible, but we must also grapple with the value question of why it is vital to do so. This next step is toward local control of energy systems which are quickly and locally built and rapidly responsive to community needs. It is another crucial chance for a more humane society in which science is sensitively applied to enhance present freedoms and future options, rather than used to enslave us under the directives of an elite few corporate technocrats and power planners. The rebirth of energy populism has only just begun.

CONSERVATION OPENS THE DOOR

Many independent researchers, universities and government agencies have already clearly detailed the first rungs on the neighborhood energy ladder. Energy conservation and weatherization—sealing up the leaks of wasted energy—not only save us money, they do three things that enable the sensible application of block-scale solar, wind and biomass energy. First, they lower individual home heating demands to the point where active solar energy systems for single homes, which aren't yet widely cost-effective, can be inexpensive if their costs are shared with other families. Second, less energy use now lengthens the time before new solar-based power plants are needed, and gives the rapidly developing community-size wind turbine, small power tower (towerlet), and urban waste bioconversion systems designers time to build demonstration plants to work the bugs out of these options. Third, with reduced energy use, smaller and therefore more nearly on-site, community-shared wind, solar and biomass energy facilities can be installed.

The amount of additional energy the new community solar energy systems have to deliver declines every time a home is insulated, a solar water heater is mounted on the roof, a house is remodeled for passive solar heating with night shutters, hot water demand is reduced with flow restrictors, and the use of a heat exchanger between used outgoing hot water and incoming hot water, and more energy-efficient, shared electrical appliances are used. This allows them to be built smaller and more humanly scaled, as well as more closely located to the actual point-of-use in the neighborhood, ready to help rooftop solar cells and solar heaters meet any larger than usual home and commercial business energy needs with a minimum of electricity and heat lost in transmission. With lower energy needs to reach the same comfort levels because we are both better managers of our presently available energy and are applying new options more efficiently, it is more likely that on only one or a few vacant lots, parking garages or commercial and business rooftops, a neighborhood energy system will be able to provide any additional energy to the residents and businesses located in the surrounding blocks.

THE TECHNICAL OPTIONS: ON THE BLOCK

Among the possibilities for block-level renewable energy systems are the following configurations:

1. *Point-of-Use Generation, Point-of-Use Storage:* This is another way of characterizing the increas-

ingly widespread, individual home or business solar energy setups, with panels or photovoltaic cells on the roof, or a wind generator in the yard (generation) where the power will be used immediately a few feet away, or shunted into a basement water or rock reservoir, or battery bank (storage), for later use in the same dwelling or office.

2. *Point-of-Use Generation, Block Storage:* Individual home or business solar, wind or wood energy sources produce power as needed, with extra energy kept and shared among those loads located all within the same block. All normal space heat, hot water and electricity needs are provided on-site by a combination of passive solar heating, passive and active solar space heating, and solar cells cooled by incoming city water which, after absorbing heat from the cells so they can "run cool" and therefore at much higher efficiency, is piped to the domestic hot water heater for further heating in the rooftop solar flatplate collectors as needed. Excess hot water and electricity from both solar cells and solar panels goes to a community battery bank and hot water storage tank from which electric backup is supplied and from which the heated water is used to supplement individual passive solar heating using hot water space heating.

3. *Block Generation, Block Storage:* Individual families and businesses pool their financial resources to buy a cooperatively owned energy system, whether solar panels, wind turbines, solar cells or methane digesters, singly or in combination, and put them on centrally located lots or rooftops, to provide all of a neighborhood's total energy needs. Hot water and electric power are also shared from a central storage point.

4. *Point-of-Use and Block Generation, Block Storage:* This is a hybrid variant in which a shared neighborhood solar power facility generates energy to supplement or back up that produced by solar systems located at the point-of-use.

The first and second options are already being rapidly installed. The Block Generation and the Point-of-Use/Block Hybrid deserve more detailed attention, for medium-scale wind turbines, urban bioconversion in community digesters and landfills, and small solar power towers are now rapidly rolling towards us on the horizon.

COMMUNITY-SCALE WIND ENERGY

Electric utilities and their sycophants in government agencies continue to propagandize us toward an energy-wastrel, all-electric, nuclear-fueled power system. Others, such as Amory Lovins, assert that we already have more than enough electricity to provide for all of our needs that really require that expensive, high-quality form of energy. Any further increase in its supply, bought at great expense in investment capital denied to other sectors of our economy, will only expand its inappropriate use for water and space heating. Both of these needs are more sensibly met with on-site solar energy at less marginal cost and less energy lost in distribution and transmission to the load.

As the debate continues on whether or not we need more electricity rather than low temperature home and industrial process heat, and transportation fuels, it is comforting to know that, in any case, local wind-electric generators can replace nuclear plants and solar power satellites in providing this versatile, high-quality energy form. If electricity is not needed, as is highly likely in most instances, then windmills may be used directly to drive, without conversion to electricity, our agricultural and industrial processes, by means of belts and pulleys, shafts and gears, hydraulic, or pneumatic (compressed air) power transmission techniques.

As to size, even large wind turbines are now turning out to be of a much more human scale than previously thought. The 200-kw DOE-NASA contractor-built wind generator located on an open gentle slope inside the city limits of Clayton, New Mexico (a small, 3000 pop. town in the northeastern part of the state) is an excellent, even aesthetic example. In addition, the fact that 10 to 15 percent of energy needs of this town of about 750 homes can be satisfied with this one wind-electric machine (the needs of 100 homes) means that only six to nine more similar-sized machines located close to downtown could supply the total municipal electric energy demand. Given the halving of energy use per home that is easily possible with a weatherization program of insulation, weatherstripping and lowered thermostats, three or four and a half more 200-kw wind turbines could do the same job.

As to cost, from Table 2, comparing large wind generators, we see that both the larger and smaller number of 200-kw machines needed could thus be replaced by not more than one or two of the larger wind turbines, and at a low, one-time capital cost per family using 8000 kwh per year amounting to 1/4 to 1/7 of the price of a new Volkswagen Rabbit, or the price of a new wood cookstove!

Supplier(s)	Location	Peak Capacity in Kilowatts	$/KW	Annual Output in Kilowatts	$ Investment for 8000 kwh/home/yr
Dakota Wind & Sun Co. (Jacobs-type with $2000 battery storage)	aver. 12 mph wind site	4	$1,675	7,200	$7,444
Same (with $750, 4kw Gemini Synchronous Inverter)	same	same	$1,363	same	$6,055
DOE-NASA contractors	Clayton, NM	200	$1,500	800,000	$3,000
WTG Energy Systems	Cuttyhunk Is., MA	200	$ 750	600,000	$2,000
DOE-NASA contractors	Boone, NC	2000	$ 750	4,800,000	$2,500
Wind Power Products Co.	San Gorgonio, CA	3000	$ 356	6,000,000	$1,424
same	same	same	$ 200 (price goal)	same	$ 800

Table 2. Wind Turbine Costs and Power Production

In fact, the delivered price of a large wind turbine from one company, Wind Power Products of Seattle, is now 1/4 to 1/2 the prices estimated for 1985 by the U.S. Department of Energy (DOE), the National Aeronautics and Space Administration (NASA) and their contractors, such as Boeing, General Electric, Westinghouse and Lockheed, and is even now approaching the oft-disbelieved $200 per kilowatt peak figure predicted by Dr. William Heronemus of the University of Massachusetts, an early, outspoken windpower advocate.

Beyond questions about size, number and price lie the number and type of jobs possible when wind machines are locally fabricated, installed, operated and maintained. Local decision-makers will find it useful to ask a potential supplier whether a particular design can be mainly locally built under license, as can the Wind Power Products turbines, to provide employment opportunities for town residents. This potential is related to the simplicity of design, sophisticated use of local materials and to the organizational style of the firm.

THE GEDSER AND THE WTG AT CUTTYHUNK

Kenneth Boulding's dictum, "It takes an awfully long time to think of the simplest things," applies very well to wind power technology. A case in point is the 200-kw Danish Gedser mill, which serves as the model for an American wind turbine with the same generator capacity, incorporating a few of the many advances in materials and control sciences that have occurred in the 20 years since the Gedser was built in 1957. The American machine is now being tested before operation on Cuttyhunk Island off the southern coast of Massachusetts by the small family firm that built it, WTG Energy Systems of Angola, New York. While America's aerospace companies complexify and overspend under lucrative government contracts, these old and new Gedsers, the result of over 80 years of wind work in Denmark and of WTG's updates, illustrate a more appropriate wind energy technology. For this penultimate mill, low cost came from elegant simplicity in materials, construction methods and controls.

179

Figure 1

Artist's rendition of Wind Power Products Co. 3-mw wind turbine to be installed in 1979 at San Gorgonio Pass for Southern California Edison Co.

WIND POWER PRODUCTS CO.—SEATTLE

In the case of wind turbines built by Charles Schachle's Wind Power Products Co. of Seattle, another innovative small business, we find two design simplifications and one canny use of local materials. The design (Fig. 1) includes, first, a generator on the ground driven by long hydraulic fluid lines to reduce weight on the tower. Second, the tower rotates into the wind as it shifts in direction so the tower's greatest structural strength is always presented to the wind's forces. Third, glue-laminated wood truss, epoxy-coated blades, less expensively fabricated in the heavily forested Pacific Northwest and therefore less costly to replace, are used instead of aluminum or fiberglass airfoils, which are 20 times the price.

Wind Power Products Co. and WTG Energy Systems are small business, family firms who approach wind energy with great eagerness for im-

provements and cost-reducing competition, while giants such as Boeing, General Electric and Lockheed cannot seem to be as inventive, inexpensive or efficient with taxpayer dollars used for federal wind energy R&D contracts to them. Howell & Todd of the Bureau of Reclamation in Denver, Colorado, in the February 24, 1978, paper, "What Size of Wind Machine Is Best for a National Electric Power Mission?" support a number of these points:

> At the present time, the number of companies actively engaged in development of wind-powered generators is very small, and all of those so engaged are industrial giants. An obvious reason is that the very large investment of overhead effort required for responding to the current targets of the U.S. Department of Energy in the range from 1.5 to 3 mw effectively bars smaller companies from participation. One important consequence is that the number of people involved in innovative thinking and experimentation on the subject of windpower is very limited, and access to the field on the part of outside inventors with perhaps radical ideas is very difficult. If the unit size of machines were small enough, and production targets were large enough, the rate of innovation and development in the field of wind power might be correspondingly greater.

and

> . . . In retrospect, it appears that much faster progress would have been made if the national effort had been planned to progress along several parallel tracks simultaneously, with alternate routes to reach important goals. It would be perhaps wise to review windpower strategy at the present time to see whether procurement of larger numbers of smaller machines would not recruit more innovators, more inventors and more cost-conscious producers into the national program, and thus follow the trail blazed by the automotive and aviation industries toward early commercialization.

Amory Lovins has suggested a further difference. Most engineers have less fun and are less inclined toward creative tinkering in large, impersonal organizations where personal responsibility is diluted and where the R&D, even when supported by every citizen's taxes, is focussed unrelentingly on producing proprietary knowledge that gives these gigantic companies a technical monopoly and a government-subsidized competitive edge against the entrance of newer, smaller firms into the market. All this obviously discourages fundamental innovation. It's a question of the appropriate organizational scale and managerial style needed to really get appropriate wind technologies.

CONSTRUCTION LEAD TIME

Wind system prices, in dollars of capital investment required for kilowatt of peak capacity output, have dropped considerably as private enterprisers have entered the large wind turbine market. As well, the amount of time required to get from the engineering drawing board to the complete, ready-to-run, power generating installation has declined as the straightforward convergence of off-the-shelf industrial components has been applied to the job rather than complex and costly aerospace corporation fabrication, as shown in Table 3.

In addition, Wind Power Products Co. has stated they can install a 3000-kw wind turbine (3 megawatt, or about 1/400 of a nuclear power plant) in one month, once they have on hand an inventory of generators and gearboxes which usually require five to six months from the placing of an order to delivery. This does not include any increases in the speed of installation due to mass production which is possible with a larger order of, say, 25, 50 or 100 wind machines. Rather, these numbers still reflect a custom-made, "one-off" ordering situation.

The importance of such short lead times in wind power generation facility installation, compared to 10-12 years for a nuclear power plant and 6-9 years for a coal plant, for regional, state and local energy demand forecasting and for local citizens' control of their own energy needs is explained in another section.

SOLAR TOWERLET

While the extreme case of applying the inappropriate nuclear model to the exploration of solar energy may be the federal government's present construction of giant power towers for solar-thermal-electricity, a smaller, possibly more sensible version is now quietly at work in Atlanta, Georgia, testing components for the larger systems. In one of the best examples ever of technology transfer, the 400 kw (thermal) "towerlet" built at the Georgia Institute of Technology (Fig. 2) represents the rediscovery of Prof. Giovanni Francia's original power plant fabricated in 1966 at Genoa, Italy.

But rather than simply its use as a test-bed, the first question should be, "Can a small power tower on a vacant lot or warehouse roof provide supplemental power to a neighborhood or industrial area?" Its best use would seem to be in providing high temperature steam for food processing such as in community canneries. If located on the boundary between a manufacturing area and a suburb, then whenever its energy wasn't needed to give superheated steam for industrial processes, it could be used to help keep neighborhood solar district hot water storage tanks "topped off." Obviously the towerlet could serve more solar homes if they were energy conserving as well, and more nearby industries if attention is paid to the efficiency gained by matching the towerlet's energy production to the end-uses actually requiring the heat ranges it produces.

And as with the non-government funded wind turbines, the straightforward design of Francia's solar steam generating plant may lend itself to community fabrication. Of special interest is a mechanical drive mount called a "kinematic motion" device or heliostat which could be made in local neighborhood machine shops. Unlike the various solid-state electronic-based sun tracking systems of large power towers for which a few large corporations supply the components, all the towerlets' heliostats are mechanically linked through a common drive shaft so that they move together by

Supplier(s)	Location	Peak Capacity in Kilowatts	Year of Installation	Construction Time
H. Morgen Smith Co.	Rutland, VT	1250	1940	24 months
DOE-NASA contractors	Plumbrook, OH	125	1976	19 months
DOE-NASA contractors	Clayton, NM	200	1977	16 months
WTG Energy Systems	Cuttyhunk Is., MA	200	1978	12 months
Wind Power Products Co.	San Gorgonio, CA	3000	1979	9 months

Table 3. Wind Turbine Construction Lead Times

a simple central clock drive mechanism to follow the sun's movement.

Figure 2

400 kw (thermal) Solar Steam Generating Plant and High Temperature Test Facility, Georgia Institute of Technology, Atlanta.

CONSTRUCTION LEAD TIME

Like community-sized wind turbines, towerlets also required less time to build than conventional thermal power plants, as shown in Table 4.

COMMUNITY METHANE

Although the production of methane gas by the anaerobic decomposition of organic matter has occurred at municipal sewage treatment plants in the U.S. and Europe since the late 1920s, this main constituent of residential and industrial "natural gas" was considered simply a useful by-product which could power sewage treatment plants until the energy crisis. Recently, much methane conversion research has focussed on building lower cost digester systems and on ways to increase the amount of gas generated per given quantity of organic matter. Table 5 compares a range of such plants.

Unfortunately, government-funded demonstrations of urban bioconversion are continuations of the "big is better" technology attitude. A prime example is the U.S. Department of Energy financed "Garbage-to-Gas" facility which became operational in May 1978 at Pompano Beach, Florida. This plant, known as RefCoM (Refuse Conversion to Methane), uses the anaerobic process to turn solid wastes and municipal sewage sludge into methane gas at the rate of 3000 cu. ft. of gas per ton of processed refuse. Remember that "processed" bit! This 1000-ton-a-day plant is estimated to produce enough gas to serve the natural gas needs of 10,000 homes, while reducing by 70 percent the volume of solid wastes and sludge. The new solid waste gasification plant is located next to a "solid waste reduction center," which shreds mixed garbage at the rate of 65 tons an hour.

Although this awareness that our cities's waste problems can become energy-producing opportunities is laudable, the tendency to use expensive,

Builder	Location	Peak Capacity in Kilowatts (thermal)	Year of Installation	Construction Time
Giovanni Francia	Marseilles, France	40	1963	12 months
Giovanni Francia	St. Ilario, Italy	60	1965	9 months
Giovanni Francia	Genoa, Italy	100	1966	5 months
Ansaldo SpA, Georgia Inst. of Technology	Atlanta, GA	400	1977	9 months (R&D project)
Ansaldo SpA	All future towerlets ordered		1978+	3 months

Table 4. Small Power Tower Output and Lead Times

182

Size and Type of Plant	Location	Daily Estimated Methane Gas Output (in cu. ft. CH₄/ c.f. of digester)	Construction Materials & Design Type	Estimated Procurement & Construction Time	Builder
Plug-flow	China	.4-.5 c.f.	concrete without plumbing	2 wks. to 3 mos; average: 1 mo.	local peasants
Batch-flow	India	.75 c.f.	Steel gas dome with plumbing	1 to 4 mos. average: 2 mo.	local peasants
75 cow Plug-flow	Ithaca, NY	1.5 c.f.		9 mo.	Dr. Bill Jewell, Cornell Univ. for U.S. DOE
75 cow Complete Mix	same	1.5 c.f.		9 mo.	same
300 cow Complete Mix	Monroe, WA	1 c.f. min. 4 c.f. max.	A.O. Smith Slurry-store tanks, complex expensive sewage treatment machinery	actual 6 mo.; 3 mo. w/standardized drawings, union construction	Ken Smith, Ecotope Group State of Wash. & & U.S. DOE
400 cow Plug-flow	North Glen, CO	2.5 c.f.		3 mo.	Energy Harvest, Inc.
1000 Household Controlled Landfill	proposed	2 c.f.	polyethylene plastic, pipe and scribbers	12 mo.	proposed (K. Smith, Dynatech R&D)
10,000 Home RefCoM "Garbage-to-Gas"	Pompano Beach, FL	3 c.f.		16 mo.	Waste Mgmt., Inc. for US DOE

Table 5. Bioconversion Plant Comparison

energy-intensive resource recovery machinery (blowers, magnets, water-washers) at the front end of this system to separate the methane producing organic materials from the inorganic metal, glass and plastics is energy-inefficient and insensitive to the greater, habit-changing potential of individual home recycling.

SOURCE SEPARATION/HOME COLLECTION

Given the increasing activity of many Americans in separating their own garbage and the many home collection recycling small businesses already existing in the neighborhoods of many communities, it would seem more sensible to encourage the personal separation of waste at its source—the home—rather than having it done for us invisibly by faraway and costly machines. Source separation continually educates people about their generation of waste, encourages them to reduce it by the real incentive of a cheaper garbage bill, and provides many times more local jobs. Recyclers who provide this pick up service charge less per month and per garbage can than the conventional "mixed-waste" garbagemen whose trucks feed the RefCoM system, because the recyclers can resell the home-separated, "pure" wastes in quantity to glass companies, paper dealers and metal fabricators. Essentially, individual householders do a bit of work in place of the breakdown-prone resource recovery machinery and hence won't be upping their local property taxes to pay the high capital costs of such equipment.

Once the urban organic waste is source-separated from inorganics which don't produce methane, the organics can be collected and delivered to a nearby neighborhood methane plant for conversion into gas or to a controlled landfill from which gas can be tapped for introduction into the community natural gas pipeline. Ken Smith, designer and builder of the 300-dairy-cow Monroe, Washington, methane plant, who now works at California's Office of Appropriate Technology, outlined this plan in a joint proposal with Dyna-Tech R&D Corporation of Cambridge, Massachusetts, to the City of Seattle in 1977. Their idea is that municipal solid waste and sewage can be combined for gas production using source separation/home collection to put the solid waste portion into a form which can be buried in a controlled landfill. Then pipes can be driven down to tap the gas generated underground. Or the digestible organics could be the raw material for town digesters located near the point-of-use since their operations are odorless, supplying neighborhood residences, small businesses and restaurants as needed.

Americans generate about three pounds of digestible matter per person per day, and each pound of matter yields five cubic feet of methane, so each person's garbage will produce about 16 cu. ft. of methane per day. Thus the potential gas production for a city of 30,000 is:

$$16 \text{ ft.}^3 \times 30,000 = 480,000 \text{ ft.}^3/\text{day of methane}$$

If half this gas is used to keep the digesters warm (and the methane bacteria alive and happily at work) at the facility, then 240,000 cu. ft. per day of gas would be available for use in town kitchens. Since cooking requirements per capita range from 7 to 10 cu. ft. of gas per day, the community's gas needs range between $(7 \text{ ft.}^3 \times 30,000)$ and $(10 \text{ ft.}^3 \times 30,000)$ = 210,000 to 300,000 cu. ft. of gas per day. This need could be more than met by the potential gas production of the city's inhabitants.

MELDING CHANGING HABITS AND A.T.

The integration of source separation/home collection of garbage and community methane conversion systems illustrates the synergistic benefits of whole systems approaches, which include not only more appropriate, human-scale technologies which can be locally situated but also an understanding of rapidly changing human behavior. Granted, we could build large resource recovery plants, but with more jobs, less expense, lower taxes, greater environmental awareness and the enhanced possibilities for local control all accruing from a slightly different option, the question must be "Will anyone, other than those who design and sell them, want them?"

CONSTRUCTION LEAD TIME: THE UNRECOGNIZED FACTOR IN LOCAL CONTROL

Since 1973 there has developed an increasingly noisy and divisive debate over whose forecast of future energy needs is correct, that of those wanting increased energy supply or of others desiring increased energy conservation, and that of those wanting more centralized electrical generation via nuclear and solar power plants or of those enamored of "Soft Path" on-site, decentralized, renewable energy systems. This often arcane and certainly confusing number game has led to even more vociferous disagreement and name-calling as the forecasts have ranged farther and farther into a future that is so obviously unknowable. It is time to re-examine the needs for such guesswork now so synonymous with public distrust. For if we are to make a societal transition to new energy forms, then we must find a starting point for agreement.

Long-range forecasts are a necessary corollary to the decade-long procurement and construction lead times needed to install large thermal power plants: 10 to 12 years for nuclear stations; 6-9 for coal plants. Yet as shown earlier, the construction times of community wind, solar or biomass are all less than two years and most under eight months. This makes long-term forecasting unnecessary even if we use appropriate energy technologies in the hundreds of kilowatts, or 1-3 mw range.

PAY AS YOU GO

Such short lead times means the capacity to locally produce electricity, heat or fuels almost simultaneous with actual, rather than highly debatable, increases in demand. In a sense the very size and local fabrication potential of the neighborhood and small town renewable energy options covered earlier makes them very close to "load following" since they can be built as needed and as we are willing to pay for them. It's "pay as we go" and not, as is now the case, paying both before and during plant construction via inflation, interest on construction work in progress, higher costs for capital, environmental standards changes during construction and rising labor rates. Lately, much has been made of the risk of under- or over-building, of not having enough energy capacity when needed or having too much lying idle while we all pay higher rates for unneeded power plants. Utilities, of course, warn us it's much safer to overbuild, yet by building as needed those smaller solar plants with much shorter lead times, consumer costs for energy are kept as low as possible. This reduces the utilities' financial risk as well as, for the possibility of under- or over-building can then approach a very near zero probability.

COMMUNITY ENERGY PSYCHOLOGY

This new "build and pay as you go" route, with the likelihood of solar energy stations in or very near the community, also reintroduces a direct and visible relationship between personal energy use and the need for more local power plants. Rather than plants located hundreds or thousands of miles away, unseen or even unknown by urban energy users, neighbors can *see* the effects of their combined power demands and what is going to be required to meet them.

One can easily imagine, as with the Clayton, New Mexico, type of highly visible, community-scale solar plant, an increase in all those subtle and direct peer group pressures to conserve energy. For if it can be seen from the picture windows of most homes that the town's windmill is not turning in the breeze, then there will be much greater personal attention to what electrical energy needs are really vital and which can be shifted to another windier time of day or to the next windy day. It will be a continuing reminder not to flick that switch so blithely—a constant environmental education based on reality. It seems likely that when such solar plants are installed, whether as towerlets or town digestors, families will want to be aware of when it is all right to use energy and that various lights, dials and indicators with local solar power production information will be located in the home or put on one cable TV channel as is now done with weather, time and local news.

LONG-RANGE MEGAWATTS VS. LOCAL KILOWATTS

The increasing inability to forecast energy needs accurately in the long run is presently due to the fact that future energy prices, once so stable or only slowly rising, are now either swiftly rising or simply unknowable and uncontrollable. The effects of new, rising energy prices or unstable, unreliable supplies on human behavior were not studied in the past and only educated guesses, based on sparse, recent data, can be made as to the likely responses of the U.S. citizen. Simultaneously with and partially fueled by the energy crisis, we have entered an era of rapidly changing personal values and behavior. Earlier, forecasting energy needs mattered little because energy prices were a rather constant and relatively small part of disposable income. Now that these costs are larger, more people are using many different ways to reduce them and strikingly new responses, such as the phenomenal growth of wood heating, are occurring. Many such individual responses amaze utility executives who thought us so dependent on electricity. It is questionable whether such near-sighted self-interest is conducive to low cost energy production for which utilities receive a public monopoly.

INCREMENTAL ENERGY

Thus, a new approach seems necessary if we are to tap the benefits of local solar, wind and methane power plants. We must begin to build in smaller, renewably-powered increments. Short lead times, no fuel costs and their amenability to local fabrication all point to the advantages of using small blocks of solar power at a time. For only such understandable, human-scale energy sources will allow us to lower our voices, agree on energy needs and proceed to realize them on a local, democratic basis to the benefit of all.

LOCAL INVESTMENT, LOCAL CONTROL

Just as the $1000-$1500 cost of a home solar water heater is within the financial means of most families, while a nuclear plant is not and requires even the largest electric utilities to float multi-billion dollar bond issues on Wall Street, community solar towerlets, methane digesters and wind generators are also much more amenable to local fundraising. Using the wind-electric example whose range of costs are shown in Table 2, a one-time investment of between $800 and $1424 allows a family to own an "8000 kwh per year share" in a 3 mw windpower plant designed to last 30 years. This means that $26.66 to $47.46 *per year* buys 8000 kwh ($2.22 to $3.95 for 666 kwh *per month*). At that rate it would be well within the ability of most households to actually buy into a community-owned electric utility in order to finance the initial cost of a wind plant installation, "owning" the asset rather than only "renting" its output. While additional money must be tacked on to these base costs for daily operation, maintenance and administrative overhead, such costs would add no more than another $1 to each family's cost of owning a piece of the wind power plant as well as the rights to a portion of its production.

EFFICIENT, PUBLICLY OWNED WIND POWER

Although community-scale wind generators are obviously a sensible investment for giant, private, investor-owned utilities, the benefits to the individual of being a member of a small, rural or municipal electric cooperative producing power from locally sited renewable energy plants are even clearer. Publicly owned utilities are simply a better deal all around. They show a national average of 30 percent lower costs per kwh than privates, exclusive of retained earnings (the equivalent of "profit" but garnered through efficiency rather than rate increase). These earnings could be used to install additional generating capacity in towerlets or methane digesters (a sort of financial "solar breeder"). Public utilities spend less on advertising, less for public relations, less for lobbying, less for local political donations, less for accounting and collecting, less for executive salaries, less for internal bureaucracy, less for unneeded and expensive office buildings, and less for private planes and other executive perks.

This efficiency applies to electric power generation also, auguring well for the economic use of solar-based power. The 1971 Federal Commerce Commission statistics report that municipal power systems have 10 percent less net electric plant per customer than investor-owned utilities, but deliver 12.2 more kilowatt-hours per customer. Delivering more electricity per customer from less plant and at lower cost is a very good indication of thrifty operation.

LOCAL CAPITAL, LOCAL MANAGEMENT

The second major benefit of public power is local control. The management of community-owned utilities is directly responsible to local residents, not to absentee stockholders. And the types, sizes and costs of sun-driven power plants already mentioned further enhance that control. It has already been shown that even for that high-quality form of ener-

gy we know as electricity, cooperative ownership of local means for its production via the earth's winds are technically available and economically within a prudent family budget, so large sources of capital external to the community need not be sought. The costs of town digesters and even less expensive controlled landfills are also amenable to local financing. However, even five years after the energy crisis, analysis is needed to determine the capital costs, local fabrication and job potential of towerlets, mainly because the U.S. Department of Energy is so influenced by the monopolistic goals of the giant private utilities that it studies and builds only five- and ten-megawatt solar power towers for our energy future.

With this ability of a community utility to avoid the high costs and control of faraway money markets, basic decisions such as plant location, rate of power plant expansion, placement of power lines, type and size of power generation, and electricity rates can continue to be made in public forums along with one's neighbors, who are also co-owners of a local solar power plant. Since the equitable sharing of power rather than private profit is their over-riding motive, community-owned systems have no built-in reason for "padding" the rate base, or for setting rates at other than a level consistent with basic costs and local policies.

THE FINAL OPTION:
DESIGNING FOR LOCAL JOBS

One of the characteristics of appropriate energy technology is the awareness and use of design options which allow local construction and increase local employment. Just as we can now build the type and size of power plant that allows rapid, local decision making on when to do so, supported by local control of the financing mechanisms that pay for it, the selection of an engineering design which lets us build a solar plant in local shops, employing local people, is also an important possibility to consider. Although many American workers and their unions now recognize the much greater job potential of conservation and solar energy compared to conventional coal and nuclear plants, we have yet to include as a major criterion for evaluating the nation's energy policies the more specific goal of local jobs in machine shops, fabrication industries and other small businesses.

There is no doubt that these new, solar-driven stations can be fabricated by giant corporations.

Indeed, neglecting the nation's pioneering solar, wind and biomass companies while giving more and larger government grants and contracts to big firms too loath to risk paying for their own R&D out of their own profits, thus driving out the innovators and inflating to unreality the final costs of renewable energy systems, is such a visible, consistent effect of federal government activity in all areas of energy hardware that it must be considered an actual policy of the U.S. Department of Energy. But this malicious neglect serves only to advertise the alternative policy: the strengthening of small towns and neighborhoods by local production with local labor of the very same community-scale solar energy systems these municipalities and neighbors can install.

Models already abound: the new attempts at solar district heating, the 50 families on Cuttyhunk Island with their own wind turbine, the engineering and operational simplicity of the Danish Gedser mill, the licensing of Wind Power Products Company's wind turbine for local fabrication around the world, the scale of the Francia solar towerlet and the potential for many of its main components to be locally made, and the decreasing design complexity and increasing use of widely available materials in town digesters and controlled landfills. As rising energy costs increase long-distance transportation costs, regional production is likely to become even more economic, ending the shipment of all but the most highly specialized parts thousands of miles, including those in neighborhood solar energy plants. Only continuing tax dollar subsidies to large aerospace corporations and giant manufacturers will prevent this natural effect of the energy crisis from rippling through our economic system and ultimately enhancing local bio-regional production and distribution.

BEYOND COMMUNITY TECHNOLOGY TO COMMUNITY DEMOCRACY: BY DESIGN

That new, technical possibilities and straightforward economics are now re-directing America's scientific and engineering knowhow toward dispersed, neighborhood energy systems should now be clear to all. But beyond the technique and the costs are the social effects which we are free to choose according to our personal and community values.

It is now within our capability to harness three main forces which can erect the powerful and humane tripod of local solar energy democracy. *Local decision making* can control when to build because construction lead times for local solar, wind and methane plants are so short as to be effectively instantaneous, making debatable long-range energy forecasts unnecessary. *Local investment* and financing can allow not only access to power at inflation-proof solar prices but actual shared, community ownership of the means of its production. *Local fabrication* and installation for local jobs can occur if design criteria are directed to give high value to local production to end unemployment.

The choice has never been clearer. If we are to realize the additional freedoms and benefits that we can confer on one another if we act together, in a sharing way, rather than the isolated few advantages that devolve on us as separate individuals, we must now work for much more than simply a transition to an all-solar society. We must look at the process and the end products of that solar world with an eye toward our nation's fundamental democratic ideals. We must now choose the scales, financing and designs for sunpower which fit into our towns and city neighborhoods and which empower them economically, socially, politically and ethically, as well as electrically. ∎

Tom Bender is one of those people who is continually pushing us to think about new ideas. It is he who has always been particularly anxious to get beyond "appropriate technology" to the joyful spiritual levels of personal and community change. His earliest perspectives as an architect/ teacher/builder/writer/prodder are in his Environmental Design Primer (6), while his latest works can be followed in Rain (48).

GETTING THROUGH THE LOOKING GLASS

By Tom Bender

OUR FIRST GLIMPSE through the looking glass into the future towards which we are headed revealed a topsy-turvy Wonderland. Things got smaller rather than larger as they got better. People played when they should have worked, and saved when they should have spent. Right was wrong and wrong was right, and the Earth Queen, not the Man-Machine, gave the last word heard. It seemed some what insane but oddly coherent. Since then we've learned the need and reasons for those different values, actions and dreams, and the insanity has been found not to be in the patterns of a stable and sustainable world. It lies, rather, in our trying to transfer the patterns that are viable in a period of rapid growth onto a period when neither growth nor its associated activities are possible or desirable.

We're moving into that Looking-Glass World— dragged in by ever-tighter pocketbooks, lured in by glimpses of a better life, pushed in by a world that no longer has either the resources or the patience to support our wasteful life-style. We may, with luck and hard work, get through the looking glass intact. We won't get through it unchanged.

Enough changes have already been made to know that our journey is possible. Communities are instituting energy-wise ordinances, planting trees and voting *not* to put in central sewers. State governments are buying windmills and banning corporate farming. People are cutting their personal energy use in half, learning to garden and cook again, insulating homes, converting to wood and solar heat, and driving less. Buildings are being dismantled rather than demolished. We're figuring out where the problems stem from and what changes must be made. Our journey is not only possible—we're well on our way.

A lot remains to be done.

We need to develop clear enough vision to unhesitatingly make the right choices in all the big and little everyday changes we must make as individuals and as a society.

We have to find the spark in ourselves to quit "letting George do it" and start doing things ourselves and getting in on the fun. The more we actually experience the benefits of doing things differently, the more clearly and certainly we can move even further.

And we need to dig in and do all the fun and dirty work of turning our institutions and rules and dreams and actions inside-out to fit our new conditions.

MIXING MEANS AND ENDS

Our goal is Life . . .
. . . experiencing its fullest and most awesome
   possibilities.
. . . healthy and continuing undiminished by our
   actions.
. . . the life of all of creation and not just ourselves.

Our goal is Liberty . . .
. . . for us and for others.
. . . for the future as well as now.
. . . within the law of nature.

Our goal is the Pursuit of Happiness . . .
. . . the happiness of being able to love people, with
   all their quirks and oddities.
. . . the happiness of being able not to *have* but to
   love our places, our things, the many-legged
   and winged and rooted peoples.
. . . being able to love the earth, the sun, the wind
   and rain, in their difficult as well as welcome
   times.
. . . being able to love ourselves.
. . . being able to love the happiness of others.

. . . being able to love the pursuit or the "doing" of
   this as much as the "having" or the end itself.
. . . being able to love these things as one—
   indivisible, not as things apart.

Keep these goals clear and before you, for they will
show you always the right path among often
seductive choices.
   Know that the goal is not wealth.
   Know that the goal is not power.
   Know that the goal is not progress.
      For the joy in these things is fleeting.

   Know that the goal is not education, or science, or
technology, or comfort, or security or health.
   These are means, at best, and things that must as
often be given up as gained to reach more important
things.

   Appropriate technology is not a worthy end in
itself. Its real value lies in creating a way of doing
things that can enable us to seek a richer and more
rewarding life.

YOU GET WHAT YOU GIVE

Changing the way we satisfy our material appetites can have a powerful effect upon our experience and reward from life. We've developed ways of feeding most of our needs and desires with minimum expenditure of effort, time, responsibility and skill of our own. That has seemed useful, on the surface of things, because it freed us to pursue and satisfy more and more of our desires.

Compared to such "quick-n-easy" ways of doing things, the alternatives often seem difficult, time consuming and archaic. Heating with a woodstove is messier and requires more attention than poking up the thermostat on your electric heat. Baking your own bread requires putting aside some time, getting your hands dirty, and cleaning up the mess. Playing with your kids takes more time and energy than plunking them down in front of the tube. A can of Campbell's is easier than homemade soup. Calling the serviceman is easier than fixing your water heater or your leaking roof.

Yet actually choosing the sometimes time-consuming and querulous alternatives reveals that they are a totally different kind of animal than the minimum-effort economic goods. For the extra time and energy put into these things is frequently not a loss, as it is seen in economic terms—it is an investment in life-enhancing experiences that are not and cannot be offered by their "labor-saving" economic replacements. The benefits and costs of different alternatives in our lives seldom lie on the same dimension. One may be economically more desirable, while another may contribute irreplaceably to other dimensions of our lives.

Why is it that a wood fire feels so comforting? Why is it that coming home to the smell of fresh-baked bread puts a warm and happy glow in our hearts? Why do we walk a little more confidently when we have just figured out and fixed our leaky plumbing?

Firelighting is a life-enhancing ritual. It reaffirms, each time, the reality of providing warmth. The fire is there. We are capable of taking care, ourselves, of getting warm. We are reminded each trip to the woodshed of the reality of our reserves. A full woodshed gives a comfort that greatly exceeds that of a full bank account. We can tell when the woodshed is full, but never know when there's *enough* in a bank account. Gas, oil or electric heat *may* be totally reliable. Likely not. But it is never fully com-

prehensible, and its heat never renews a sense of assurance.

Similarly, most ways of providing for our needs that require a great deal of ourselves to be put into the work return a great deal to us. The more we know how to do things ourselves, the better our crap-detectors are for anything which is poorly done. In many things, such as home-prepared foods, doing it yourself is the *only* true assurance of quality or the only way to have something done well. And the love that often goes into and shines through such things is rarely put into working for a wage to buy the store-bought alternative. Quite the contrary, most wage-earning, consumer-buying situations are infested with deceit, frustration, anger, distrust and dissatisfaction as a result of their breaking a task into competing and self-serving roles.

In providing for each of our needs there is unique potential for enhancing our lives—for improving quality on the material level, for subtly and repeatedly filling deep-buried needs for assurance and confidence in our own abilities, and for deepening our grasp and confidence in the things upon which we draw for our existence. In their repetition, these actions become life-giving rituals—giving fullness, happiness and assurance to our lives in measure to our putting ourselves into them.

ADDING DIMENSION TO OUR LIVES

There is a subtle yet powerful poison and opiate in the ancient blood of dinosaurs upon which our society feasts that has robbed us of our vitality and wholeness. The poison is called plenty. Dinosaur blood, along with the bones of the past in the riches of an unexploited continent, has overwhelmed our society with material things, with the kinds of satisfactions they bring us, with the means—economics —by which they are measured and manipulated, and with the means—intellectual—by which they are brought into being. These things, with their apparent satisfaction, apparent power and apparent bountifulness, have lured us into throwing our energies into their further development and procurement at the expense of all other aspects of our lives. The products of our opiate have long passed the point of diminished returns, and the supply of the opiate itself is diminishing. We are left with an awareness that we've lost something, but we can't remember what.

We can't remember the heights and depths of experiences that give intensity, strength, beauty, character and harmony to our lives. What we can't remember are the life-enhancing relationships of love, trust, respect, caring, sharing and giving that forge the enduring and rewarding links between us and others. We've lost our link with the rest of creation that gives joy, meaning and perspective to our lives. We can't remember what enables us to approach everything we do, say, think, feel or dream in ways that turn them into life-enhancing, spirit-sustaining and character-developing celebrations. We can't remember how to reach the integrative, jump-making, deep-experiencing part of our minds that assure us of the rightness or wrongness of a course of action. These things are the connectors of life rather than the dissectors, the cement rather than the bricks with which we build a rewarding life, the integrative rather than the disintegrative factors in our relationships. They are powerful forces, and beside them the experience of a merely material and intellectual life pales. Together, and in balance, they can give a quality and substance to our lives that has long been lost.

Our wealth has influenced how we do things, in ways that block their contribution to non-economic dimensions of our lives. In changing how we do things—in removing the cushions of wealth and technology—we can reconnect ourselves more directly to these other dimensions.

Changing the way we respond to loss or tragedy is one way that can open for us new and valuable levels of inner resources. Unsought, unprepared for, and usually inexplicable, loss of life, jobs, friends, home or possessions all plunge deep into our soul. Our society tries to shield us and cushion us from the pain and realities of these inevitable traumas of life, yet it is the difficult but direct dealing with them ourselves that gives us depth of comprehension and compassion. Vigils with the dead in earlier times acted to give focus to the grief, memories and loss, and to help us bring out and deal with the feelings and emotions our present customs try to distract us from. Wakes acted to share and celebrate what the survivors had shared and been given during the life of the dead person. These events give us a clearer and stronger sense of the wonder and fragility of life and a greater impetus to savor each moment of it. These are painful and draining processes at the time, but leave us with ways through which our inner strengths can pour forth and through which we can more easily gain

access to that strength, wisdom, love and understanding when we have need for it.

We instinctively turn to our wealth to find material ways of comforting us and cushioning us from these experiences just as we have used that wealth to cushion us from other stresses of life. In so doing, we increase our real loss rather than gain from it.

Knowing these other dimensions changes our attitude towards our material possessions, just as loss of possessions from fire, theft or other means leads us to appreciate more the value of valued things. Learning how much things mean to us in being links to family, friends, special times, or being special in themselves, makes us realize how little, in comparison, the economic value of things is. We discover that we only want things around us that can give more than economic value. We also learn why it is possible to give away but impossible to sell something that we love very much. The giving adds value to it for both the other person and yourself, while the value of the other meanings remains intact. Selling it reduces it to only its economic value.

In an economics-centered world, we are constantly pushed to accept that there is only economic value to things. Special meaning of special things is never given an economic value (though in part it could) because the cost would be too great (an admission of the true value of valuing things!) The economic system has to force down such values to the point where they don't interfere with "progress" towards its underlying goals. . . . "We realize you love your family farm, but we want to put a highway here, and you can buy 'similar' land elsewhere for $xxx an acre." . . . "I know these things are priceless, but you have to set a value on them for insurance purposes. How large a premium are you willing to pay?" . . . "We must optimize the value of this National Forest. It's worth $xxx an acre for timber production. What is the recreational value worth?"

These heights and depths of experience that lie outside of economics are invaluable to our lives. The discipline to direct and limit our material impulses strengthens in turn with the rewards it gives us in these other dimensions.

JOBS, HUMBUG!

Do we need jobs? Should we *want* jobs? Are jobs a desirable part of life, liberty and our pursuit of happiness? Or are jobs a dead-end trap in themselves? Having a job means working for someone else. It means getting locked into one specific pattern of providing for our needs that may be particularly socially destructive, emotionally unsatisfying, and wasteful. Jobs may be a good option in some situations, but we usually ignore the alternatives, which are often better.

Jobs, on the whole, give us more than a paycheck. They give us a lot of satisfactions or dissatisfactions, security or ulcers. They lock us into a cash economy, specialization and not taking care of our own needs. They split us into opposing and self-serving roles— producer and consumer, employer and employee, management and labor. Jobs, like store-bought bread, rarely come in different sized slices—it's full- or no-time. Thought of providing jobs to counter unemployment easily falls into who can "provide" jobs—large institutions—rather than the less centralized and less manipulable alternative of people creating work for themselves. Having jobs can be as much a basic cause of social and economic problems as a solution to them.

There are alternatives to jobs—alternatives that have far different social as well as economic consequences. We can work for ourselves (self-employment). We can provide for our own needs (self-reliance). Or we can demand less and work less (self-restraint).

It's amazing how much energy in a job goes into taking advantage of others or resisting being taken advantage of. Figuring out how to pay people less, or get more work out of them. Figuring out how to stretch coffee breaks and lunch, get to work late, get more sick days, to look like you're busy. Bucking for promotion. Taking credit for the work of people below you to get a raise. Passing on blame. Being angry at management for its stupidity or meanness. Unionizing to fight for fair wages and working conditions. Hassling with the *union* management. Feeling, rightly, that you probably don't matter.

Alternatives to jobs free us from huge amounts of this bad personal energy by avoiding the division of work into those conflicting and exploitive elements that are inherent in job relationships.

Self-employment avoids the division of interest between worker and management. You've no one to get mad at but yourself, and there's no profit in trying to pull one over on yourself. Self-reliance goes even further in eliminating the split between the producer and consumer. You know what you're getting, and the price is right, because the price is whether you're willing to put in the necessary work to do it! Self-restraint takes us another important step towards surmounting divisive conflicts of

interest. Demanding less and thereby avoiding unnecessary production and consumption lessens our demands upon our resources and each other, lessening the conflict between us and our grandchildren for limited resources and between our greed and the health of our surroundings.

The "job" relationship that is so common in industrial society puts us into a real double-bind. We're supposed to be efficient, productive and work for the interests of our employer for a certain period, then abruptly turn around and be profligate consumers, buying for whim, vanity, luxury and prestige. At the same time we're trapped by an effective divide and conquer strategy by commercial interests. By dividing us alternatively between working and consuming roles—neither of which we control or in which we can bargain as equal partners —we get milked coming and going. As individual consumers we have little power to ensure true quality or value in what we purchase. The profit lies in deceiving the consumer. As individual employees we have little power to gain fair working conditions or a fair division of income between our work and the inputs of management. As consumers or employees banded together into groups or unions with enough power to balance the corporations, we suddenly find ourselves with another large and powerful bureaucracy to deal with.

Ways of meeting our needs more directly and more simply also meet them more satisfyingly. Self-confidence grows with self-reliance. The more we're responsible for satisfying our own needs, the less we're trapped in the frustrations, anger and distrust that fill the marketplace and the workplace. Self-reliance minimizes a person's cash income and therefore taxes. Real needs can often be better known and more effectively met by the person having those needs.

Development of more employment-intensive and capital and energy-saving production processes are essential in the changes our society must make. They do not, however, ask the more basic question of whether the production was necessary in the first place, or if the work is necessary for the production. More durable goods or more restraint in our material demands coupled with better sharing of necessary work remaining in necessary production would offer greater benefits than unnecessary production or work for the sake of its employment or income potential alone.

Appropriate technologies offer valuable employment, capital and energy benefits, and greater op-

192

tions in scale and location of work. Even more, though, they allow these basic changes from strife-ridden adversary work to more unitary patterns of the rewards they can bring.

CONSENSUS DEMOCRACY

The ways by which we take action are frequently as important as the actions themselves. If we want a society that lives equality rather than talks it, where the thoughts and feelings of every person are listened to and incorporated into decisions and actions, and which can attain and maintain a harmonious and deep-rooted relation with the rest of Nature, we need to change our ways of making public decisions. Majority-rule democracy will not work.

Majority-rule processes have been widely used in a period when we have given tacit consent to rapid change—specifically because they allow us to ignore and override the desires of significant numbers of people who would never consent to the changes proposed or the way they are done. A self-serving minority able to swing the vote of people not understanding the implications of the action being taken or who have been led to believe it might benefit them has repeatedly overridden the real interests of the community by this process.

Urban growth serves as a well-documented example. The high costs of growth to the community as a whole have been repeatedly shown. The percent who profit are small, and their identity predictable. Yet the local businesspeople give support in hope of expanding their own operations—not realizing that growth brings in more and larger competitors and they are as likely to lose as to gain from the process. "Everyone" appears to have opportunity to profit in real estate exchanges, yet the homeowners forget they must purchase another home in the same inflated market, and the real profits always seem to go to the same people with inside information, contacts, credit and knowledge of the game.

Majority-rule voting results too frequently in energy being put into obtaining a majority rather than in listening to and coming to terms with the real and important feelings of the minority. It creates a divisiveness in carrying out decisions between those victoriously carrying out their wishes and the losers grudgingly accepting the imposition of the decisions upon their lives. Majority-rule tends towards what is popular or easy rather than what is

right, and gives little power to the always necessary voices of dissension. It gives an illusion of strength and permanence to decisions that belies the always shifting feelings of a community. It responds to the interests of power, not people.

We tend to consider majority rule as the only workable form of democratic decisionmaking. Yet one alternative in particular, *consensus*, is far more democratic and far more respectful of the community as a whole and responsive to it. In consensus, decisions must be acceptable to everyone. To us, used to approving actions which *aren't* acceptable to everyone, that sounds impossible. It isn't. Native American, Chinese, Quaker, and many Japanese cultures have long employed consensus. Its use in one strongly polarized U.S. government planning committee was amazingly effective at getting people's feelings clearly articulated, at involving everyone in an effort to find a workable solution, and in leaving everyone with a sense of commitment to make the decisions work.

At first impression, consensus is unwieldy and slow compared to simple voting. But it results in real differences being worked out rather than being swept under the table and pulls the energies of the whole group together behind a decision rather than causing the obstruction, indifference and uncommitted assistance more common with voted decisions. It requires a group or community to deal with its real problems on a more honest, open and direct level, which in itself is a major improvement over how we do things now. Consensus also differs from majority rule in that it is basically stabilizing. Failure to agree upon a new action doesn't mean inaction —it means merely that things go on as they have unless or until full agreement on change is reached.

In reality, majority rule represents a false economy. The minutes saved in reaching a "decision" are more than lost in implementation, in the anger, frustration and rejection felt by losing voters, and in the repeated cropping up of the unresolved differences at every possible opportunity. A real community needs the solidarity of shared respect for each other and of shared and accepted direction which can only emerge from consensus kinds of processes.

THE ECONOMICS OF DEMOCRACY

No system in which some people have a vastly greater ability to impose their will upon others can be a democracy. Regardless of its trappings, it becomes a dictatorship of power—either economic or political.

Our present economics encourages immense concentrations of economic wealth—not only in degree but in orders of magnitude—that result in massive imbalance of political power. These differences in wealth and power are not the normal ones that arise from differences between those who work hard and those who play hard, between those who spend and those who save, or between those with differing abilities. They arise not as a result of a person's own efforts but from exploitation of the work or results of work of others. They are the result of an economic philosophy and structure that encourages manipulation of the whole system of laws and regulations governing economic activity to increase the wealth and power of the already wealthy.

Differences in personal income don't reveal the true imbalance that exists. The surplus income of the wealthy accumulates. The expenditures of the poor do not. Statistics on personal *wealth* distribution in this country show greatly more disparity than those for income, but even they barely hint at the pyramiding of power that our economic system grants to wealth. Examples of that pyramiding of power occur blatantly in places such as real estate, where a person's investment of wealth or equity is multiplied by everyone else's small savings in form of a mortgage loan. The people using or renting the property are the ones who actually pay off that mortgage, but their payments act to increase the equity of the investor rather than giving themselves a share of ownership in the property.

All that, though, is small game compared to the cornering of power that results from the rules governing corporate behavior we have allowed to be set up. Control of a corporation requires only a small proportion (sometimes as little as 3 percent) of voting stock—as long as the rest is broken up into smaller ownership blocks and the other owners don't gang up on you. And if the corporation you control controls other corporations through similar stock ownership, etc., your power quickly pyramids into corporate holding companies and holding companies that control holding companies. Mutually beneficial friendships among the wealthy and powerful give such individuals even further power through positions on interlocking directorships of corporations.

The result is a vast concentration of power among a very few people—power over jobs and wages, power to decide what and how things are produced, sold and consumed. Power to determine what prices farmers receive and consumers pay and how much is siphoned off in the middle. Power to get self-serving legislation, to finance politicians, and to control the press. The power, in sum, to exert vast control over other people's lives. The further such concentration proceeds, the more difficult it is to challenge and control it.

The direct measures that need to be taken to regain power over our own lives are fairly simple. Concentration and forms of wealth and power that are unnecessary to our collective well-being cannot be allowed. That means measures such as prohibiting *any* corporate operation in many areas, such as farming. Several states have already imposed such a ban and seen its beneficial results. Corporate ownership of stock of other corporations should not be allowed, nor should a person be permitted to act as a director of more than one company. Multiplant and conglomerate firms provide us no economic benefit and should not be permitted. Renters should receive ownership equity equal to the amount of their rent that goes to mortgage repayment.

A great many more specific changes are needed to reverse the myriad self-serving regulations that the wealthy have fostered upon us over the recent years of our worship of bigness. But they need most importantly to be rooted in a clear and decisive sense of an economics that serves democracy rather than one that serves power.

Our economic system has focussed almost totally on specialization and trade or exchange rather than upon self-reliance—either personal or geographic. We usually work at one specific job and trade the income from that for the housing, food, medical care, entertainment and transportation that we could to a large degree provide for ourselves. Similarly, regions and nations specialize in manufacturing, bananas, coffee, electronics, etc., and trade those specialties for other things—but in markets that rarely provide equitable exchange.

In the abstract, such an exchange economy—whether barter- or money-based—appears to have the advantage of greater expertise and efficiency of production. That may or may not be true or a benefit when all costs are considered. But it does have one fundamental and fatal flaw compared to a self-reliant economy where a person, community or region provides for most of its own needs. It is manipulable.

It is manipulable by a powerful buyer or seller that can impose unfair conditions and prices. It is Safeway, not farmers or consumers, that sets food prices. It is manipulable by the U.S. gunboats in the harbor that really set terms when a small nation signs trade agreements with the U.S. It is manipulable by the total control banks have over use of our savings. It is manipulable by the power large institutions have over the lives of their employees. It is manipulable by the taxation that supports centralized services while also forcing everyone into a cash/exchange economy controlled by and serving the wealthy. It is manipulable by our ignorance, as specialists, of the true cost, value and nature of things outside our own specialty.

To the degree that we provide for our own needs, we escape the exploitive activities of such systems. To the degree that we exchange our services for those of others, under conditions or through mechanisms over which we don't have a fair share of control, we contribute with our own sweat to the further imbalance of power and wealth in our society. If a family grows much of its own food, it has less need to purchase food from commercial food systems controlled by others, and less need to work for money in situations frequently dominated by others in order to pay for that food. If a region grows most of its own food, recycles iron and steel in small plants within the region, minimizes its need for fuels imported from other areas, and prohibits outside ownership of local enterprises—to that degree it escapes participation in national and international markets and systems controlled and exploited by others. If a community controls and pays for its own sewage treatment, education and health needs, it escapes the strings attached to federal dollars.

It becomes quite clear whose interests are served by an exchange-based economy vs. a self-reliant one by looking at the process by which our exchange system has spread throughout the world. Japan did not clamor and beg to trade with us in the 18th century—we forced it with our military presence. It was Western industrialists looking for markets in China that forced the Opium Wars and eventual establishment of a trade economy. American smallhold farmers did not choose to stop self-reliant homesteading and move to the city—they were forced out by taxes to support urban and governmental services, price squeezes caused by monopoly control of equipment, and food markets and governmental policies favoring large farms.

It is not surprising that alternatives to an exchange economy have been systematically ignored. It is not surprising that our first (and second) rate engineering has gone into the technology of specialization and exchange rather than self-reliance, giving them the appearance of inherent superiority. It is not surprising that in spite of the great wealth of our society the actual non-inflated income and well-being of most people has steadily decreased in recent years.

It appears now that a true comparison of self-reliant vs. exchange economies would yield some interesting surprises. The political and organizational impacts of economies dominated by one or the other of these systems are fairly obvious and so fundamental as to overshadow any but the most extreme differences in costs of goods and services. But the economic impacts appear to be quite the opposite of what the promoters of "economies of scale" have led us to believe. The recent focus of competent effort into the development and application of small-scale technologies that both have sustainable resource demands and sustainable social and cultural impact has considerably changed the picture.

We now know a great deal about the operation, cost and impacts of such things as compost toilets, solar space and water heating, home food production, small-scale industrial plants, owner-building or co-op banking. The tradeoffs compared to their conventional alternatives are enlightening. Satisfactory compost toilet/greywater systems are available at costs between one and ten percent of a household's share of the cost of a sewer system. Owner-building that can avoid financing of a home can provide housing at approximately one-tenth the ultimate cost of a financed, commercially built dwelling. Decentralized renewable energy production provides substantial employment benefits to a community as well as economic savings. Locally owned and operated restaurants, stores and industry recycle money within the community in contrast to franchises or chainstores whose outside corporate ownership siphons up to 60 percent of their cash flow out of a community.

The picture that is coming together suggests that a composite economy based upon a substantial degree of self-reliance of regions, communities and individuals can offer a direct *order of magnitude* reduction of real economic costs in addition to other benefits such as a more satisfying life and preservation of democratic distribution of power and wealth in our society. We need careful consideration of such economies—examining household expenditures and alternative means of meeting the needs they correspond to; examining alternatives to present community and governmental services; examining major regional industries, employment and needs, and other means of dealing with them. We'll likely find that a lot of our customary ways of thinking about things will change. We may discover that increased property values do not mean increased personal worth but only increased property costs to everyone and increased benefits only to municipal tax departments and a small number of speculators. And we'll start changing things that encourage such increases. We'll probably discover that "making a profit" can either mean making a living or making a killing, and find ways to discriminate, to nurture the former and to control the latter. We'll likely find that there is real truth behind the rural and small town distrust of city money. And we'll probably discover that self-reliant economies involve work whose effect, both on the worker and in what is produced, is directly understood and valued— rather than aimless, meaningless paper-shuffling and assembly line drudgery!

If the monetary savings of self-reliance turn out to be anywhere near what they look to be, a whole new world of options opens to us. Options to be more relaxed in our work and to restore the satisfactions of work well done and work done in more rewarding ways that we have given up in our forced pursuit of efficiency. Options to press less heavily on others and on our resources. Options to put our effort into creating a beautiful and satisfying world rather than a wealthy, efficient or powerful one. Options to create a new Golden Age with the freedom and energy to pursue the harmonious development of all our capabilities—that to love or to make music as well as that to make money.

New dreams are possible. The Golden Age of almost every society in history has occurred *not* when all a culture's energies were focussed on increasing its wealth and power, but rather when the attainable limits of those dreams were reached and people realized that such goals had not left them with the quality, beauty or personal happiness they had envisioned. Freeing the vast resources of society that are now channeled into things that are no longer attainable makes new visions possible and attainable. Once we realize that greatness is not

achievable through vast expenditures of resources but requires the development and refinement of our own personal abilities, we discover that our present wealth is more than adequate to achieve an equitable and golden age for the whole world.

Moving away from our recent way of life and the incredible wealth of energy and resources that have been consumed in making it possible means saying goodbye to an unprecedented and possibly unrepeatable era in the history of our planet. But that is true, in different ways, of many other eras—both in the past and into which we now move. We should not look back with regret at its passing, but with thanks. For a dream has always recurred—a dream of great wealth that could allow everyone to live as kings and which would make a better or more perfect world. We have had that wealth, and have lived as kings. It is not, in all, a good life or a better world.

Knowing that is both vital and unobtainable in any way other than we have gotten it. Our true wealth lies within us, within the myriad other things that share and make up our world, and within the wonder of the strange and beautiful universe of which we are all part. It lies in responding to, not subjugating these things. In that is the wellspring of a truly great era. ∎

THE STAGE IS bare now. We are between theories. We are in the last period of the fossil fuel era—and the so-called nuclear era is already aborting.

What we miss is something as simple as a vision of how we will live in the future.

No one sees the future; we have no clear images —as a culture, as a nation, as the Western world.
When the stage is empty there is unprecedented opportunity.

When the stage of the future is unoccupied, when there is not one strong vision of which we are all in the process of working out, we don't have to fight *against* either the established vision or the rebels. There is no enemy. The empty stage is the *rarest* of opportunities. Then build a future, make it work, and let the world steal it.

Ty Cashman
From RAIN, 1977 (48)

RESOURCES

(1) AMBIO—JOURNAL OF THE ROYAL SWEDISH ACADEMY OF SCIENCES, $29/year (individual) from: P.O. Box 142, Boston, MA 02113.

(2) ARCHITECTURAL DESIGN, $48/year from: 7/8 Holland, London, W8, England.

(3) Baer, Steve, SILLY PRESIDENT CARTER, *Rain: Journal of Appropriate Technology*, April, 1978, $1 from: 2270 N.W. Irving, Portland, OR 97210.

(4) Baldwin, J. & Stewart Brand, SOFT TECH, 1978, $5 from: Box 428, Sausalito, CA 94965.

(5) Barnet, Richard & Ronald Muller, GLOBAL REACH: THE POWER OF MULTINATIONAL COR-PORATIONS, 1974, $4.95 from: Simon & Schuster (a Gulf & Western Company!), 630 Fifth Ave., New York, NY 10020.

(6) Bender, Tom, ENVIRONMENTAL DESIGN PRIMER, 1973, $5.95 from: Schocken Books, 200 Madison Ave., New York, NY 10016.

(7) Bender, Tom, SHARING SMALLER PIES, 1975, $2 from: *Rain*, 2270 N.W. Irving, Portland, OR 97210.

(8) Berry, Wendell, THE LONG-LEGGED HOUSE, 1971, Bantam Books, Out of Print.

(9) Berry, Wendell, THE UNSETTLING OF AMERICA: CULTURE AND AGRICULTURE, 1977, $9.95 from Sierra Club Books, 530 Bush St., San Francisco, CA 94108.

(10) Bookchin, Murray, TOWARDS A LIBERATORY TECHNOLOGY, *Post-Scarcity Anarchism*, 1971, $3.95 from: Ramparts, P.O. Box 10128, Palo Alto, CA 94303.

(11) BRIARPATCH BOOK, 1978, $8 from: New Glide Publications, 330 Ellis St., San Francisco, CA 94102.

(12) BRIARPATCH REVIEW, quarterly, $5/year from: 330 Ellis St., San Francisco, CA 94102.

(13) Burns, Scott, HOUSEHOLD ECONOMY: IT'S SHAPE, ORIGINS AND FUTURE, 1975, $4.95 from: Beacon Hill Press, 25 Beacon St., Boston, MA 02108.

(14) Clark, Wilson, ENERGY FOR SURVIVAL, 1974, $4.95 from: Doubleday & Co., 245 Park Ave., New York, NY 10017.

(15) Clark, Wilson, BIG AND/OR LITTLE? SEARCH IS ON FOR THE RIGHT TECHNOLOGY, *Smithsonian Magazine*, July, 1976, $10/year from: Smithsonian Institution, Washington, DC 20560.

(16) CO-EVOLUTION QUARTERLY, $12/year from: Box 428, Sausalito, CA 94965.

(17) Coomaraswamy, Ananda K., THE INDIAN CRAFTSMAN, 1909, Probsthain & Company, London, Out of Print.

(18) Daly, Herman, TOWARD A STEADY-STATE ECONOMY, 1973, $3.95 from: W. H. Freeman & Company, 660 Market St., San Francisco, CA 94104

(19) Darrow, Ken & Rick Pam, APPROPRIATE TECHNOLOGY SOURCEBOOK, 1976, $4 from: Volunteers in Asia, Box 4543, Stanford, CA 94305.

(20) Douglas, J. Sholto & Robert A. deJ. Hart, FOREST FARMING, 1976, $8.95 from: Rodale Press, Emmaus, PA 18049.

(21) ENVIRONMENT ACTION BULLETIN, Rodale Press, Emmaus, PA 18049. No Longer Published.

(22) Fathy, Hassan, ARCHITECTURE FOR THE POOR, 1975, $5.95 from: University of Chicago Press, 5801 Ellis Ave., Chicago, IL 60637

(23) FOREIGN AFFAIRS QUARTERLY, $10/year from: Council on Foreign Relations, 58 East 68th St., New York, NY 10021.

(24) Gregg, Richard, VOLUNTARY SIMPLICITY, *Manas*, Sept. 4 & 11, 1974, $10/year from: P.O. Box 32112, El Sereno Station, Los Angeles, CA 90032.

(25) Hess, Karl and David Morris, NEIGHBORHOOD POWER, 1975, $3.45 from: Beacon Press, 25 Beacon St., Boston, MA 02108.

(26) Howard, Sir Albert, THE SOIL AND HEALTH, 1947, $3.95 from: Schocken Books, 200 Madison Ave., New York, NY 10016.

(27) Illich, Ivan, TOOLS FOR CONVIVIALITY, 1973, $1.25 from: Harper & Row, 10 East 53rd, New York, NY 10022.

(28) Illich, Ivan, TOWARD A HISTORY OF NEEDS, 1977, $7.95 from: Pantheon Books, 201 East 50th, New York, NY 10022.

(29) King, F. H., FARMERS OF FORTY CENTURIES, 1911, $8.95 from: Rodale Press, Emmaus, PA 18049.

(30) Kohr, Leopold, OVERDEVELOPED NATIONS: DISECONOMIES OF SCALE, 1977, $9.95 from: Schocken Books, 200 Madison Ave., New York, NY 10016.

(31) Lappé, Frances Moore, DIET FOR A SMALL PLANET, 1975, $1.95 from: Ballentine Books, 201 East 50th, New York, NY 10022.

(32) Lappé, Frances Moore and Joseph Collins, FOOD FIRST, 1977, $7.95 from: Institute for Food Development Policy, 2588 Mission St., San Francisco, CA 94110.

(33) Lotka, A. J., CONTRIBUTION TO THE ENERGETICS OF EVOLUTION, *Proceedings of the National Academy of Science, Vol. 8, 1922.*

(34) Lovins, Amory, *ENERGY STRATEGY: THE ROAD NOT TAKEN, 1975, 50¢* from: Friends of the Earth, 124 Spear St., San Francisco, CA 94105.

(35) Lovins, Amory, SOFT ENERGY PATHS: TOWARD A DURABLE PEACE, 1977, $6.95 from: Friends of the Earth.

(36) Lovins, Amory, WORLD ENERGY STRATEGIES, 1975, $4.95 from: Friends of the Earth.

(37) Lovins, Amory and John Price, NONNUCLEAR FUTURES: THE CASE FOR AN ETHICAL ENERGY STRATEGY, 1975, $5.95 from: Friends of the Earth.

(38) MANAS, weekly, $10/year from: P.O. Box 32112, El Sereno Station, Los Angeles, CA 90032.

(39) Marsh, George P., MAN AND NATURE, 1878, Harvard University Press, Out of Print.

(40) Merrill, Richard and Thomas Gage, ENERGY PRIMER, 1978 (revised edition), $7.95 from: Delta Books 1 Dag Hammarskjold Plaza, New York, NY 10017.

(41) Mill, John Stuart, THE PRINCIPLES OF POLITICAL ECONOMY, Vol. II, 1857, John W. Parker & Son, Out of Print.

(42) THE NEW FARM, new monthly from: Rodale Press, Emmaus, PA 18049.

(43) NEW MEXICO SOLAR ENERGY SOCIETY NEWSLETTER, $10 individual membership, from: P.O. Box 2004, Santa Fe, NM 87501.

(44) North, Michael, TIME RUNNING OUT? THE BEST OF RESURGENCE, 1976, $5.95 from Universe Books, 381 Park Ave., New York, NY 10016.

(45) Odum, Howard T., ENVIRONMENT, POWER & SOCIETY, 1971, $6.50 from: Wiley Interscience, 605 Third Ave., New York, NY 10016.

(46) Odum, Howard T. and Elizabeth, ENERGY BASIS FOR MAN & NATURE, 1976, $7.95 from: McGraw Hill, 1221 Avenue of the Americas, New York, NY 10036.

(47) QUEST, monthly, $12/year from: Ambassador International Cultural Foundation, 300 West Green St., Pasadena, CA 91129.

(48) RAIN: JOURNAL OF APPROPRIATE TECHNOLOGY, monthly, $15/year from: 2270 N.W. Irving, Portland, OR 97210.

(49) RAINBOOK: RESOURCES IN APPROPRIATE TECHNOLOGY, 1977, $7.95 from: 2270 N.W. Irving, Portland, OR 97210.

(50) RESURGENCE, monthly $10/year ($15 airmail) from: Pentre Ifan, Felindre, Crymych, Dyfed, Wales, U.K.

(51) Schatz, Joel and Tom Bender, INDEPENDENCE?, 1974, Oregon Office of Energy Research & Planning, Out of Print.

(52) Schumacher, E. F., SMALL IS BEAUTIFUL, 1973, $2.95 from: Harper & Row, 10 East 53rd St., New York, NY 10022.

(53) Schumacher, E. F., TECHNOLOGY AND POLITICAL CHANGE, reprinted in *Rain*, Dec., 1976 & Jan., 1977, $1 each from: 2270 N.W. Irving, Portland, OR 97210.

(54) SELF-RELIANCE, bi-monthly, $6/year from: 1717 18th Ave., N.W., Washington, DC 20009.

(55) SMITHSONIAN MAGAZINE, $10/year from: Smithsonian Institution, Washington, DC 20560.

(56) Snyder, Gary, FOUR CHANGES, *Turtle Island*, 1974, $1.95 from: New Directions, 333 Avenue of the Americas, New York, NY 10014.

(57) Stein, Barry, SIZE, EFFICIENCY & COMMUNITY ENTERPRISE, 1974, $5 from: Center for Community Economic Development, 639 Massachusetts Ave., Cambridge, MA 02139.

(58) Todd, Nancy Jack, THE BOOK OF THE NEW ALCHEMISTS, 1977, $6.95 from: E. P. Dutton, 201 Park Ave. So., New York, NY 10003.

(59) VanderRyn, Sim, APPROPRIATE TECHNOLOGY AND STATE GOVERNMENT, 1975, 75¢ from: Office of Appropriate Technology, 1530 10th St., Sacramento, CA 95814.

(60) Wells, Malcolm, UNDERGROUND DESIGNS, 1977, $6 from: P.O. Box 1149, Brewster, MA 02631.

(61) WHOLE EARTH CATALOG, 1971, $5, and WHOLE EARTH EPILOG, 1974, $4 from: Box 428, Sausalito, CA 94965.

(62) Wint, Guy, ASIA: A HANDBOOK, 1966, Anthony Bond, Ltd., London

ORGANIZATIONS

(63) ACORN/Midwest Energy Alternatives Network, Governor's State University, Park Forest South, IL 60466.

(64) ALTERNATIVE ENERGY RESOURCES ORGANIZATION (AERO), 435 Stapleton Bldg., Billings, MT 59101.

(65) CENTER FOR COMMUNITY ECONOMIC DEVELOPMENT, 639 Massachusetts Ave., Cambridge, MA 02139.

(66) FARALLONES INSTITUTE, 15290 Coleman Valley Rd., Occidental, CA 95465.

(67) INSTITUTE FOR LOCAL SELF-RELIANCE, 1717 18th St., N.W., Washington, DC 20009.

(68) NATIONAL CENTER FOR APPROPRIATE TECHNOLOGY (NCAT), P.O. Box 3939, Butte, MT 59701.

(69) NEW ALCHEMY INSTITUTE, P.O. Box 432, Woods Hole, MA 02543

(70) OFFICE OF APPROPRIATE TECHNOLOGY (OAT), State of California, 1530 10th St., Sacramento, CA 95814.

(71) OFFICE OF SMALL SCALE TECHNOLOGY, U.S. Dept. of Energy, Washington, D.C. 20545.

(72) VOLUNTEERS IN ASIA (VIA), Box 4543, Stanford, CA 94305.

(73) VOLUNTEERS IN TECHNICAL ASSISTANCE (VITA), 3706 Rhode Island Ave., Mt. Ranier, MD 20822.

INDEX